CHAPMAN & HALL/CRC
Monographs and Surveys in
Pure and Applied Mathematics   107

DIRECT AND INDIRECT BOUNDARY INTEGRAL EQUATION METHODS

CHRISTIAN CONSTANDA

CHAPMAN & HALL/CRC
Monographs and Surveys in Pure and Applied Mathematics

Main Editors

H. Brezis, *Université de Paris*
R.G. Douglas, *Texas A&M University*
A. Jeffrey, *University of Newcastle upon Tyne (Founding Editor)*

Editorial Board

H. Amann, *University of Zürich*
R. Aris, *University of Minnesota*
G.I. Barenblatt, *University of Cambridge*
H. Begehr, *Freie Universität Berlin*
P. Bullen, *University of British Columbia*
R.J. Elliott, *University of Alberta*
R.P. Gilbert, *University of Delaware*
R. Glowinski, *University of Houston*
D. Jerison, *Massachusetts Institute of Technology*
K. Kirchgässner, *Universität Stuttgart*
B. Lawson, *State University of New York*
B. Moodie, *University of Alberta*
S. Mori, *Kyoto University*
L.E. Payne, *Cornell University*
D.B. Pearson, *University of Hull*
I. Raeburn, *University of Newcastle*
G.F. Roach, *University of Strathclyde*
I. Stakgold, *University of Delaware*
W.A. Strauss, *Brown University*
J. van der Hoek, *University of Adelaide*

CHAPMAN & HALL/CRC
Monographs and Surveys in
Pure and Applied Mathematics   **107**

DIRECT AND INDIRECT

BOUNDARY INTEGRAL

EQUATION METHODS

CHRISTIAN CONSTANDA

CHAPMAN & HALL/CRC
Boca Raton   London   New York   Washington, D.C.

Library of Congress Cataloging-in-Publication Data

Constanda, C. (Christian)
    Direct and indirect boundary integral equation methods / Christian Constanda.
        p. cm. — (Monographs and surveys in pure and applied mathematics ; 107)
    Includes bibliographical references.
    ISBN 0-8493-0639-6 (alk. paper)
    1. Boundary element methods. I. Title. II. Series: Chapman & Hall/CRC monographs and surveys in pure and applied mathematics ; 107.
TA347.B69C68   1999
515'.35—dc21   99-41796
                                                                                  CIP

This book contains information obtained from authentic and highly regarded sources. Reprinted material is quoted with permission, and sources are indicated. A wide variety of references are listed. Reasonable efforts have been made to publish reliable data and information, but the author and the publisher cannot assume responsibility for the validity of all materials or for the consequences of their use.

Neither this book nor any part may be reproduced or transmitted in any form or by any means, electronic or mechanical, including photocopying, microfilming, and recording, or by any information storage or retrieval system, without prior permission in writing from the publisher.

The consent of CRC Press LLC does not extend to copying for general distribution, for promotion, for creating new works, or for resale. Specific permission must be obtained in writing from CRC Press LLC for such copying.

Direct all inquiries to CRC Press LLC, 2000 N.W. Corporate Blvd., Boca Raton, Florida 33431.

Trademark Notice: Product or corporate names may be trademarks or registered trademarks, and are used only for identification and explanation, without intent to infringe.

© 2000 by Chapman & Hall/CRC

No claim to original U.S. Government works
International Standard Book Number 0-8493-0639-6
Library of Congress Card Number 99-41796
Printed in the United States of America   1 2 3 4 5 6 7 8 9 0
Printed on acid-free paper

For Lia

Contents

Preface

Chapter 1. The Laplace Equation   1

  1.1. Notation and prerequisites   1
  1.2. The fundamental boundary value problems   7
  1.3. Green's formulae   9
  1.4. Uniqueness theorems   12
  1.5. The harmonic potentials   14
  1.6. Properties of the boundary operators   17
  1.7. The classical indirect method   24
  1.8. The alternative indirect method   29
  1.9. The modified indirect method   31
  1.10. The refined indirect method   37
  1.11. The direct method   40
  1.12. The substitute direct method   48

Chapter 2. Plane Strain   54

  2.1. Notation and prerequisites   54
  2.2. The fundamental boundary value problems   58
  2.3. The Betti and Somigliana formulae   61
  2.4. Uniqueness theorems   66
  2.5. The elastic potentials   67
  2.6. Properties of the boundary operators   71
  2.7. The classical indirect method   85
  2.8. The alternative indirect method   92
  2.9. The modified indirect method   95
  2.10. The refined indirect method   102
  2.11. The direct method   105
  2.12. The substitute direct method   114

Chapter 3. Bending of Elastic Plates   121

  3.1. Notation and prerequisites   121
  3.2. The fundamental boundary value problems   126

3.3. The Betti and Somigliana formulae	129
3.4. Uniqueness theorems	135
3.5. The plate potentials	136
3.6. Properties of the boundary operators	139
3.7. Boundary integral equation methods	143

Chapter 4. Which Method?   145

4.1. Notation and prerequisites	145
4.2. Connections between the indirect methods	146
4.3. Connections between the direct and indirect methods	150
4.4. Overall view and conclusions	152

Appendix   154

A1. Geometry of the boundary curve	154
A2. Properties of the boundary layer	157
A3. Integrals with singular kernels	164
A4. Potential-type functions	176
A5. Other potential-type functions	182
A6. Complex singular kernels	190
A7. Singular integral equations	195

References   200

Preface

Boundary integral equation methods are among the most powerful and elegant techniques developed by analysts for solving elliptic boundary value problems. Applied to a wide variety of situations in mathematical physics and mechanics, these procedures have the great advantage that they deliver the solution in closed form, which is very helpful for numerical computation.

It may be surprising, but the construction of boundary integral methods in two-dimensional cases is, as a rule, more difficult than in higher dimensions. This anomaly stems from the growth of the relevant fundamental solutions in the far field and their variable sign in the domain where they are defined, which may give rise to "pathologies". Because of these drawbacks, such problems have largely been neglected in the literature, being passed over in articles and books with the general remark that their treatment is similar to the handling of their three-dimensional counterparts.

This book gives a full yet succint description of four indirect and two direct methods in application to the interior and exterior Dirichlet, Neumann and Robin problems for the Laplace equation (Chapter 1) and the systems governing plane strain (Chapter 2) and bending of thin elastic plates (Chapter 3). In Chapters 1 and 2 the discussion is complete, to allow the reader to see both the common features and the differences arising in the construction of integral representations for the solutions of a single equation and a system of equations. By contrast, Chapter 3 goes only as far as setting up the necessary tools (plate potentials, boundary integral operators, characteristic boundary matrices) and leaves out the solvability proofs, which, with the obvious changes, would simply duplicate those given for plane strain. In Chapter 4 the various direct and indirect techniques introduced earlier are compared and contrasted in order to establish which one is the most suitable for numerical approximations.

The highly technical details that are essential in the study of the behaviour and properties of potentials near the boundary—what one might call "microanalysis"—are gathered together in a rather substantial Appendix at the end; these details are for the benefit of the reader who wants to know not only how, but also why the methods work.

Parts of the book are based on results published previously in [7]–[15].

It is true to say that many authors have a favourite place where they go to recharge their depleted batteries and regain a sense of purpose and enthusiasm when their candle is burning low. For me this is the University of Tulsa. Bill Coberly, the head of the Department of Mathematical and Computer Sciences, and his colleagues there have always been most generous and welcoming, creating the kind of atmosphere in which one feels almost compelled to write books. I thank them all for their help and friendship.

I would also like to express my thanks to my friends at the University of Strathclyde for their solid support and encouragement, and to the staff of CRC Press, who have handled the process of publication with skill and patience, from the commissioning stage to production.

But the greatest debt of gratitude I owe is without question to my wife. Although not a mathematician, she understands that mathematics is a passion which feeds on itself and sometimes temporarily locks the practitioner in a universe of his own. Like the true professional that she is, during such intervals she has always stood by me and engineered my rescue with sound advice and good humour. This book could not have been written without her indirect but invaluable assistance.

July 1999 *Christian Constanda*

Chapter 1
The Laplace Equation

1.1. Notation and prerequisites

Let S^+ be a finite domain in $\mathbb{R}^2$ bounded by a simple closed C^2-curve ∂S, and let $S^- = \mathbb{R}^2 \setminus (S^+ \cup \partial S)$. We denote by $x = (x_1, x_2)$ a generic point in $\mathbb{R}^2$ referred to a Cartesian system of coordinates with the origin in S^+, and by $\nu(x)$ the outward (with respect to S^+) unit normal at $x \in \partial S$. We use the notation

$$\partial_{\nu(x)} = \partial/\partial\nu(x), \quad \partial_{s(x)} = \partial/\partial s(x)$$

for the normal and tangential derivatives at $x \in \partial S$, respectively.

If u is a function defined in a domain that includes ∂S, then we denote the restriction of u to ∂S by $u|_{\partial S}$. However, for the normal derivative of u on ∂S we simply write $\partial_\nu u$, since ∂_ν is itself a boundary operator.

1.1. Definition. A function f defined on $\bar{S}^+ = S^+ \cup \partial S$ is called *Hölder continuous* (with index $\alpha \in (0, 1]$) on $\bar{S}^+$ if

$$|f(x) - f(y)| \leq c|x - y|^\alpha \quad \text{for all } x, y \in \bar{S}^+, \tag{1.1}$$

where $c = \text{const} > 0$ is independent of x and y and

$$|x - y| = \left[(x_1 - y_1)^2 + (x_2 - y_2)^2\right]^{1/2}.$$

If f is defined on $\bar{S}^- = S^- \cup \partial S$, then (1.1) must hold on every bounded subdomain of $\bar{S}^-$. In this case the constant c may vary with the subdomain.

We denote by $C^{0,\alpha}(\bar{S}^+)$ the vector space of all Hölder continuous (with index α) functions on $\bar{S}^+$, and by $C^{1,\alpha}(\bar{S}^+)$ the subspace of $C^1(\bar{S}^+)$ of all differentiable functions whose first order derivatives belong to $C^{0,\alpha}(\bar{S}^+)$. The spaces $C^{0,\alpha}(\bar{S}^-)$, $C^{1,\alpha}(\bar{S}^-)$ and $C^{0,\alpha}(\partial S)$, $C^{1,\alpha}(\partial S)$ are defined similarly.

Here we present a brief review of some concepts and results that are necessary in what follows. A complete discussion and full proofs can be found in the Appendix.

Let $\kappa(x)$ be the algebraic value of the curvature at $x \in \partial S$. Since ∂S is a C^2-curve, we can define

$$\kappa_0 = \max_{x \in \partial S} |\kappa(x)|.$$

We set
$$q = 4\max\{2\kappa_0, \kappa_0^2 l\},$$
where $l = |\partial S|$ is the length of ∂S, and introduce the boundary layer
$$S_0 = \{x \in \mathbb{R}^2 : x = \xi + \sigma\nu(\xi),\ \xi \in \partial S,\ |\sigma| \le \tfrac{1}{4}\min\{\tfrac{1}{2},\ \kappa_0^{-1},\ (2q)^{-1}\}\}.$$

1.2. Definition. A two-point function $k(x,y)$ defined and continuous for all $x \in S_0$ ($x \in \partial S$) and $y \in \partial S$, $x \ne y$, is called a γ-*singular kernel in* S_0 (*on* ∂S), $\gamma \in [0,1]$, if there is an $a = \text{const} > 0$, which may depend on ∂S, such that for all $x \in S_0$ ($x \in \partial S$), $x \ne y$,
$$|k(x,y)| \le a|x-y|^{-\gamma}.$$
If, in addition,
$$|k(x,y) - k(x',y)| \le a|x-x'|\,|x-y|^{-\gamma-1}$$
for all x, $x' \in S_0$ (x, $x' \in \partial S$) and $y \in \partial S$ satisfying
$$0 < |x-x'| < \tfrac{1}{2}|x-y|,$$
then $k(x,y)$ is called a *proper γ-singular kernel in* S_0 (*on* ∂S).

We extend this definition to two-point matrix functions by requiring each component of the matrix to satisfy the necessary properties.

1.3. Theorem. (i) *If $\varphi \in C(\partial S)$ and $k(x,y)$ is continuous in $S_0 \times \partial S$ and such that $\mathrm{grad}(x)k(x,y)$ is a proper γ-singular kernel in S_0, $\gamma \in [0,1)$, then the function defined by*
$$(v^a\varphi)(x) = \int_{\partial S} k(x,y)\varphi(y)\,ds(y), \quad x \in S_0,$$
belongs to $C^{1,\alpha}(S_0)$, with $\alpha = 1-\gamma$ for $\gamma \in (0,1)$ and any $\alpha \in (0,1)$ for $\gamma = 0$.

(ii) *If $\varphi \in C(\partial S)$ and $k(x,y)$ is continuous on $\partial S \times \partial S$ and such that $(\partial_{s(x)}k)(x,y)$ is a proper γ-singular kernel on ∂S, $\gamma \in [0,1)$, then the function defined by*
$$(v_0^a\varphi)(x) = \int_{\partial S} k(x,y)\varphi(y)\,ds(y), \quad x \in \partial S,$$
belongs to $C^{1,\alpha}(\partial S)$, with $\alpha = 1-\gamma$ for $\gamma \in (0,1)$ and any $\alpha \in (0,1)$ for $\gamma = 0$.

It is clear that if the smoothness properties of k with respect to x extend beyond S_0 in Theorem 1.3(i), then so do those of $v^a\varphi$.

1.4. Theorem. (i) *If $\varphi \in C(\partial S)$, then the functions defined by*

$$(v\varphi)(x) = \int_{\partial S} (\ln|x-y|)\varphi(y)\,ds(y), \quad x \in \mathbb{R}^2,$$

$$(v^b_{\beta\delta}\varphi)(x) = \int_{\partial S} (x_\beta - y_\beta)(x_\delta - y_\delta)|x-y|^{-2}\varphi(y)\,ds(y), \quad x \in \mathbb{R}^2,$$

$$(v^c_\beta\varphi)(x) = \int_{\partial S} \left[\partial_{s(y)}\big((x_\beta - y_\beta)\ln|x-y|\big)\right]\varphi(y)\,ds(y), \quad x \in \mathbb{R}^2,$$

$$(v^d_\beta\varphi)(x) = \int_{\partial S} \left[\partial_{\nu(y)}\big((x_\beta - y_\beta)\ln|x-y|\big)\right]\varphi(y)\,ds(y), \quad x \in \mathbb{R}^2,$$

belong to $C^{0,\alpha}(\mathbb{R}^2)$ for any $\alpha \in (0,1)$.

(ii) *If $\varphi \in C^{0,\alpha}(\partial S)$, $\alpha \in (0,1)$, then the function defined by*

$$(v^e_{\beta\delta}\varphi)(x) = \int_{\partial S} \left[\partial_{s(y)}\big((x_\beta - y_\beta)(x_\delta - y_\delta)|x-y|^{-2}\big)\right]\varphi(y)\,ds(y), \quad x \in \mathbb{R}^2,$$

belongs to $C^{0,\alpha}(\mathbb{R}^2)$.

(iii) *If $\varphi \in C^{0,\alpha}(\partial S)$, $\alpha \in (0,1)$, then the functions $v_0\varphi$, $v^b_{\beta\delta 0}\varphi$, $v^c_{\beta 0}\varphi$, $v^d_{\beta 0}\varphi$ and $v^e_{\beta\delta 0}\varphi$ defined as in (i) and (ii) but with $x \in \partial S$ belong to $C^{1,\alpha}(\partial S)$.*

(iv) *If $\varphi \in C^{0,\alpha}(\partial S)$, $\alpha \in (0,1)$, then the restrictions of the function defined by*

$$(w\varphi)(x) = \int_{\partial S} (\partial_{\nu(y)} \ln|x-y|)\varphi(y)\,ds(y), \quad x \in S^+ \cup S^-,$$

to S^+ and S^- are $C^{0,\alpha}$-extendable to $\bar{S}^+$ and $\bar{S}^-$, respectively.

(v) *If $\varphi \in C^{0,\alpha}(\partial S)$, $\alpha \in (0,1)$, then the function $w_0\varphi$ defined as in (iv) but with $x \in \partial S$ exists in the sense of principal value uniformly for all $x \in \partial S$ and belongs to $C^{1,\alpha}(\partial S)$.*

(vi) *If $\varphi \in C^{0,\alpha}(\partial S)$, $\alpha \in (0,1)$, then the function defined by*

$$(v^f\varphi)(x) = \int_{\partial S} (\partial_{s(y)} \ln|x-y|)\varphi(y)\,ds(y), \quad x \in S^+ \cup S^-,$$

is $C^{0,\alpha}$-extendable to $\mathbb{R}^2$.

(vii) *If $\varphi \in C^{0,\alpha}(\partial S)$, $\alpha \in (0,1)$, then the function $v^f_0\varphi$ defined as in (vi) but with $x \in \partial S$ exists as principal value uniformly for all $x \in \partial S$ and belongs to $C^{0,\alpha}(\partial S)$.*

(viii) *If $\varphi \in C^{1,\alpha}(\partial S)$, $\alpha \in (0,1)$, then the function $v^f_0\varphi$ defined in (vii) belongs to $C^{1,\alpha}(\partial S)$.*

(ix) If $\varphi \in C^{0,\alpha}(\partial S)$, $\alpha \in (0,1)$, then the restrictions to S^+ and S^- of the function $v\varphi$ defined in (i) are $C^{1,\alpha}$-extendable to $\bar{S}^+$ and $\bar{S}^-$, respectively.

(x) If $\varphi \in C^{1,\alpha}(\partial S)$, $\alpha \in (0,1)$, then the restrictions to S^+ and S^- of the function $w\varphi$ defined in (iv) are $C^{1,\alpha}$-extendable to $\bar{S}^+$ and $\bar{S}^-$, respectively.

We now discuss briefly a few concepts of functional analysis, which will enable us to solve the boundary value problems stated later in §1.2. Once again, fuller details and proofs can be found in the Appendix.

Let z and ζ be the complex numbers corresponding (in the usual way) to the points x and y in $\mathbb{R}^2$.

1.5. Theorem. $C^{0,\alpha}(\partial S)$, $\alpha \in (0,1)$, is a Banach space with respect to the norm defined by

$$\|\varphi\|_\alpha = \|\varphi\|_\infty + |\varphi|_\alpha,$$

where

$$\|\varphi\|_\infty = \sup_{z \in \partial S} |\varphi(z)|,$$

$$|\varphi|_\alpha = \sup_{\substack{z,\zeta \in \partial S \\ z \neq \zeta}} \frac{|\varphi(z) - \varphi(\zeta)|}{|z - \zeta|^\alpha}.$$

1.6. Definition. Let X and Y be normed spaces. A linear operator $K : X \to Y$ is called *compact* if it maps any bounded set in X into a relatively compact set in Y (that is, a set in which every sequence contains a convergent subsequence).

1.7. Theorem. If $k(z,\zeta)$ is a proper γ-singular kernel on ∂S, $\gamma \in [0,1)$, then the operator K defined by

$$(K\varphi)(z) = \int_{\partial S} k(z,\zeta)\varphi(\zeta)\,d\zeta, \quad z \in \partial S,$$

is a compact operator from $C^{0,\alpha}(\partial S)$ to $C^{0,\alpha}(\partial S)$, with $\alpha = 1 - \gamma$ for $\gamma \in (0,1)$ and any $\alpha \in (0,1)$ for $\gamma = 0$.

1.8. Remark. In the Appendix it is shown that the kernels of the integral operators v_0, v_0^a, $v_{\beta\delta}^b$, $v_{\beta0}^c$, $v_{\beta0}^d$, $v_{\beta\delta}^e$ and w_0 in Theorems 1.3(ii) and 1.4(iii),(v) are proper γ-singular on ∂S, $\gamma \in [0,1)$; consequently, by Theorem 1.7, the corresponding operators are compact. Such operators are called *weakly singular*.

1.9. Definition. Let X and Y be two vector spaces over $\mathbb{C}$. A mapping

$$(\cdot,\cdot): X \times Y \to \mathbb{C}$$

is called a *non-degenerate bilinear form* if
 (i) for any $\varphi \in X$, $\varphi \neq 0$, there is $\psi \in Y$ such that $(\varphi,\psi) \neq 0$, and for any $\psi \in Y$, $\psi \neq 0$, there is $\varphi \in X$ such that $(\varphi,\psi) \neq 0$;
 (ii) for any $\varphi_1, \varphi_2, \varphi \in X$, $\psi_1, \psi_2, \psi \in Y$ and $\alpha_1, \alpha_2 \in \mathbb{C}$

$$(\alpha_1\varphi_1 + \alpha_2\varphi_2, \psi) = \alpha_1(\varphi_1,\psi) + \alpha_2(\varphi_2,\psi),$$
$$(\varphi, \alpha_1\psi_1 + \alpha_2\psi_2) = \alpha_1(\varphi,\psi_1) + \alpha_2(\varphi,\psi_2).$$

1.10. Definition. By a *dual system* (X,Y) we understand a pair of normed spaces X and Y together with a non-degenerate bilinear form $(\cdot,\cdot): X \times Y \to \mathbb{C}$.

1.11. Definition. Let (X,Y) be a dual system with bilinear form $(\cdot,\cdot)$. Two operators $K: X \to X$ and $K^*: Y \to Y$ are called *adjoint* if for all $\varphi \in X$ and $\psi \in Y$

$$(K\varphi, \psi) = (\varphi, K^*\psi).$$

1.12. Remark. It is shown without difficulty [2] that if an operator $K: X \to X$ has an adjoint $K^*: Y \to Y$ in a dual system (X,Y), then K^* is unique, and both K and K^* are linear.

1.13. Definition. Let (X,Y) be a dual system with bilinear form $(\cdot,\cdot)$, $K: X \to X$ an operator that has a (unique) adjoint $K^*: Y \to Y$, I the identity operator (which, for simplicity, is denoted by the same symbol regardless of the space where it acts), and $\omega \in \mathbb{C}$, $\omega \neq 0$, and consider the equations

$$(K - \omega I)\varphi = f, \quad f \in X, \tag{K}$$
$$(K^* - \omega I)\psi = g, \quad g \in Y, \tag{K^*}$$

together with their homogeneous versions (K_0) and (K_0^*). We say that *the Fredholm Alternative holds for K in (X,Y)* if either
 (i) both (K_0) and (K_0^*) have only the zero solution, in which case (K) and (K^*) have unique solutions for any $f \in X$ and $g \in Y$, respectively, or
 (ii) (K_0) and (K_0^*) have finitely many linearly independent solutions $\{\varphi_1, \ldots, \varphi_n\}$ and $\{\psi_1, \ldots, \psi_n\}$, in which case (K) and (K^*) are solvable, respectively, if and only if

$$(f, \psi_i) = 0, \quad (g, \varphi_i) = 0, \quad i = 1, \ldots, n.$$

1.14. Theorem. *If (X, Y) is a dual system and $K : X \to X$ a compact linear operator that has a (unique) compact adjoint $K^* : Y \to Y$, then the Fredholm Alternative holds for K in (X, Y).*

The proof of this assertion can be found, for example, in the monograph [2].

1.15. Corollary. *If K is a weakly singular integral operator (see Remark 1.8), then the Fredholm Alternative holds for K in the dual system $(C^{0,\alpha}(\partial S), C^{0,\alpha}(\partial S))$, $\alpha \in (0, 1)$, with the bilinear form*

$$(\varphi, \psi) = \int_{\partial S} \varphi(\zeta) \psi(\zeta) \, d\zeta, \quad \varphi, \psi \in C^{0,\alpha}(\partial S). \tag{1.2}$$

The Fredholm Alternative does not hold in general for operators with proper 1-singular kernels on ∂S, called *(strongly) singular operators*. However, there is a class of such operators for which the assertion remains true.

1.16. Definition. An operator $K : C^{0,\alpha}(\partial S) \to C^{0,\alpha}(\partial S)$, $\alpha \in (0, 1)$, is called *α-regular singular* if it is defined by an expression of the form

$$(K\varphi)(z) = \int_{\partial S} \hat{k}(z, \zeta)(\zeta - z)^{-1} \varphi(\zeta) \, d\zeta, \quad z \in \partial S,$$

where $\hat{k}(z, \zeta)$ belongs to $C^{0,\alpha}(\partial S)$ with respect to each variable, uniformly relative to the other one, and satisfies the inequality

$$|\hat{k}(z, \zeta) - \hat{k}(z', \zeta)| \leq c |z - z'| |z - \zeta|^{\alpha - 1}, \quad c = \text{const} > 0,$$

for all z, z', $\zeta \in \partial S$ such that $0 < |z - z'| < \frac{1}{2} |z - \zeta|$. (The value of $\hat{k}(z, \zeta)$ at $z = \zeta$ may also be understood in the sense of continuous extension.)

1.17. Definition. Consider the equation (K) with $X = Y = C^{0,\alpha}(\partial S)$, $\alpha \in (0, 1)$, and the bilinear form (1.2). Also, suppose that K is an α-regular singular operator and that

$$-\omega \pm \pi i \hat{k}(z, z) \neq 0 \quad \text{for all } z \in \partial S. \tag{1.3}$$

The number

$$\varrho = \frac{1}{2\pi} \left[\arg \frac{-\omega - \pi i \hat{k}(z, z)}{-\omega + \pi i \hat{k}(z, z)} \right]_{\partial S},$$

where $\big[\theta(z)\big]_{\partial S}$ denotes the change in $\theta(z)$ as z traverses ∂S once anticlockwise, is called the *index* of the operator K (or of the equation (K)).

If (K) is a matrix equation where φ and f are $(n \times 1)$-vectors, then $\hat{k} \in \mathcal{M}_{n \times n}$, the modulus on the left-hand side of the inequality in Definition 1.16 is understood to apply to each component of $\hat{k}$, condition (1.3) becomes

$$\det\left[-\omega E_n \pm \pi i \hat{k}(z,z)\right] \neq 0 \quad \text{for all } z \in \partial S$$

and the index is defined as [20]

$$\varrho = \frac{1}{2\pi}\left[\arg\frac{\det\left(-\omega E_n - \pi i \hat{k}(z,z)\right)}{\det\left(-\omega E_n + \pi i \hat{k}(z,z)\right)}\right]_{\partial S}, \tag{1.4}$$

where E_n is the identity $(n \times n)$-matrix.

1.18. Remarks. (i) It can easily be verified that the index of the adjoint operator K^* (or of the adjoint equation (K^*)) is $-\varrho$.

(ii) Any weakly singular operator (see Remark 1.8) is α-regular singular for any $\alpha \in (0,1)$ and satisfies $\hat{k}(z,z) = 0$. The proof can be found in the Appendix.

1.19. Theorem. *If K is an α-regular singular operator, $\alpha \in (0,1)$, of index zero, then the Fredholm Alternative holds for K in the dual system $\left(C^{0,\alpha}(\partial S), C^{0,\alpha}(\partial S)\right)$ with the bilinear form* (1.2).

1.20. Remark. In [20] it is shown that if the Fredholm Alternative holds for the operator K in the complex system $(C^{0,\alpha}(\partial S), C^{0,\alpha}(\partial S))$, then it also holds for it in the real system $(C^{0,\alpha}(\partial S), C^{0,\alpha}(\partial S))$, provided that we restrict ourselves to real solutions of (K) and (K^*), where K^* is the real adjoint of K.

1.2. The fundamental boundary value problems

We denote by $\mathcal{A}$ the vector space of functions u in S^- which, in terms of polar coordinates r, θ, satisfy the asymptotic relations

$$u(r,\theta) = O(r^{-1}), \quad (\partial_r u)(r,\theta) = O(r^{-2}) \quad \text{as } r \to \infty, \tag{1.5}$$

uniformly with respect to θ, where $\partial_r = \partial/\partial r$. We also define $\mathcal{A}^* = \mathcal{A} \oplus \mathbb{R}$. In what follows, any function written symbolically as $\mathcal{U}^{\mathcal{A}}$ or $\mathcal{U}^{\mathcal{A}^*}$ belongs to $\mathcal{A}$ or $\mathcal{A}^*$, respectively.

1.21. Definition. Let $\mathcal{P}, \mathcal{Q}, \mathcal{K}, \mathcal{R}, \mathcal{S}, \mathcal{L}, \sigma \in C(\partial S)$ be prescribed functions, with $\sigma > 0$. The interior and exterior Dirichlet, Neumann and Robin problems are formulated as follows.

(D$^+$) Find $u \in C^2(S^+) \cap C^1(\bar{S}^+)$ such that
$$\Delta u(x) = 0, \quad x \in S^+, \quad \text{and} \quad u(x) = \mathcal{P}(x), \quad x \in \partial S.$$

(N$^+$) Find $u \in C^2(S^+) \cap C^1(\bar{S}^+)$ such that
$$\Delta u(x) = 0, \quad x \in S^+, \quad \text{and} \quad (\partial_\nu u)(x) = \mathcal{Q}(x), \quad x \in \partial S.$$

(R$^+$) Find $u \in C^2(S^+) \cap C^1(\bar{S}^+)$ such that
$$\Delta u(x) = 0, \quad x \in S^+, \quad \text{and} \quad (\partial_\nu u + \sigma u)(x) = \mathcal{K}(x), \quad x \in \partial S.$$

(D$^-$) Find $u \in C^2(S^-) \cap C^1(\bar{S}^-) \cap \mathcal{A}^*$ such that
$$\Delta u(x) = 0, \quad x \in S^-, \quad \text{and} \quad u(x) = \mathcal{R}(x), \quad x \in \partial S.$$

(N$^-$) Find $u \in C^2(S^-) \cap C^1(\bar{S}^-) \cap \mathcal{A}$ such that
$$\Delta u(x) = 0, \quad x \in S^-, \quad \text{and} \quad (\partial_\nu u)(x) = \mathcal{S}(x), \quad x \in \partial S.$$

(R$^-$) Find $u \in C^2(S^-) \cap C^1(\bar{S}^-) \cap \mathcal{A}^*$ such that
$$\Delta u(x) = 0, \quad x \in S^-, \quad \text{and} \quad (\partial_\nu u - \sigma u)(x) = \mathcal{L}(x), \quad x \in \partial S.$$

1.22. Remark. Considering the Laplace equation does not restrict the generality of the problems since the non-homogeneous (Poisson) equation can be reduced to the homogeneous one by means of a particular solution (constructed, for example, in terms of an area potential [16]).

1.23. Definition. A function u satisfying any of the above six sets of equations and conditions is called a *regular solution* of the corresponding boundary value problem. Since there is no danger of ambiguity, from now on we use the term *solution*, for short.

In the case of (N$^\pm$) the boundary data must satisfy a necessary condition of solvability.

1.24. Theorem. *If* (N$^+$) *and* (N$^-$) *are solvable, then*
$$\int_{\partial S} \mathcal{Q}\,ds = 0, \quad \int_{\partial S} \mathcal{S}\,ds = 0,$$

respectively.

Proof. The divergence theorem states that, for a smooth function F,

$$\int_{S^+} \operatorname{div} F \, da = \int_{\partial S} F \cdot \nu \, ds.$$

If u is a solution of (N^+), then, by the divergence theorem,

$$0 = \int_{S^+} \Delta u \, da = \int_{S^+} \operatorname{div}(\operatorname{grad} u) \, da = \int_{\partial S} (\operatorname{grad} u) \cdot \nu \, ds = \int_{\partial S} \partial_\nu u \, ds = \int_{\partial S} \mathcal{Q} \, ds.$$

If u is a solution of (N^-), then we consider a disk K_R of radius R sufficiently large so that $\bar{S}^+$ lies strictly inside K_R, and apply the divergence theorem in $\bar{S}^- \cap \bar{K}_R$. Since the boundary of this domain consists of ∂S and the circle ∂K_R, and since the outward normal on ∂S in this case is directed into S^+, we find that

$$0 = \int_{S^- \cap K_R} \Delta u \, da = \left(-\int_{\partial S} + \int_{\partial K_R} \right) \partial_\nu u \, ds. \tag{1.6}$$

In terms of polar coordinates with the pole at the centre of K_R, the second integral on the right-hand side above can be written as

$$\int_0^{2\pi} (\partial_R u) R \, d\theta.$$

Since $u \in \mathcal{A}$, from (1.5) we deduce that the integrand is $O(R^{-1})$, so, as $R \to \infty$, (1.6) yields the desired equality for $\mathcal{S}$.

1.25. Definition. A solution u of the Laplace equation $\Delta u = 0$ in S^+ (S^-) is called a *harmonic function in S^+ (S^-)*.

1.3. Green's formulae

In the theory of boundary value problems for the Laplace and other elliptic equations a fundamental role is played by the so-called Green's identities and Green's representation formulae.

1.26. Theorem. *If $u, v \in C^2(S^+) \cap C^1(\bar{S}^+)$, then*

$$\int_{S^+} u \Delta v \, da + \int_{S^+} (\operatorname{grad} u) \cdot (\operatorname{grad} v) \, da = \int_{\partial S} u \partial_\nu v \, ds, \tag{1.7}$$

$$\int_{S^+} (u \Delta v - v \Delta u) \, da = \int_{\partial S} (u \partial_\nu v - v \partial_\nu u) \, ds. \tag{1.8}$$

Proof. Since

$$\operatorname{div}(u \operatorname{grad} v) = (\operatorname{grad} u) \cdot (\operatorname{grad} v) + u \Delta v,$$

$$u(\operatorname{grad} v) \cdot \nu = u \partial_\nu v,$$

we apply the divergence theorem to $u(\operatorname{grad} v)$ and establish the first identity. Interchanging u and v in it, we obtain

$$\int_{S^+} [v\Delta u + (\operatorname{grad} v) \cdot (\operatorname{grad} u)] \, da = \int_{\partial S} v \partial_\nu u \, ds,$$

and subtracting this from the first identity, we arrive at Green's second identity.

1.27. Corollary. (i) *If u is harmonic in S^+, then*

$$\int_{S^+} |\operatorname{grad} u|^2 \, da = \int_{\partial S} u \partial_\nu u \, ds.$$

(ii) *If $u \in \mathcal{A}^*$ is harmonic in S^-, then*

$$\int_{S^-} |\operatorname{grad} u|^2 \, da = -\int_{\partial S} u \partial_\nu u \, ds.$$

Proof. (i) This assertion follows immediately from (1.7) with $v = u$.

(ii) Consider a disk K_R of radius R sufficiently large so that $\bar{S}^+$ lies strictly inside K_R. Applying (i) in $\bar{S}^- \cap \bar{K}_R$ and noting that the outward normal on ∂S is now directed into S^+, we obtain

$$\int_{S^- \cap K_R} |\operatorname{grad} u|^2 \, da = \left(-\int_{\partial S} + \int_{\partial K_R} \right) u \partial_\nu u \, ds.$$

Since $u \in \mathcal{A}^*$, it follows that, by (1.5), the second integral on the right-hand side is of the form

$$\int_0^{2\pi} [O(R^{-1}) + \operatorname{const}] O(R^{-2}) R \, d\theta;$$

hence, this integral tends to zero as $R \to \infty$, which leads to the desired formula.

It is well known that the two-point function

$$g(x, y) = -(2\pi)^{-1} \ln |x - y| = g(y, x)$$

is a fundamental solution for the operator $-\Delta$; that is,

$$\Delta\big((2\pi)^{-1}\ln|x-y|\big) = \delta(|x-y|),$$

where δ is Dirac's delta distribution. Using $g(x,y)$, we can now derive Green's representation formulae for harmonic functions in $S^{\pm}$.

1.28. Theorem. (i) *If u is harmonic in S^+, then*

$$\int_{\partial S} [g(x,y)(\partial_\nu u)(y) - u(y)(\partial_{\nu(y)}g)(x,y)]\,ds(y) = \begin{cases} u(x) & x \in S^+, \\ \tfrac{1}{2}u(x) & x \in \partial S, \\ 0 & x \in S^-. \end{cases}$$

(ii) *If $u \in \mathcal{A}$ is harmonic in S^-, then*

$$-\int_{\partial S} [g(x,y)(\partial_\nu u)(y) - u(y)(\partial_{\nu(y)}g)(x,y)]\,ds(y) = \begin{cases} 0 & x \in S^+, \\ \tfrac{1}{2}u(x) & x \in \partial S, \\ u(x) & x \in S^-. \end{cases}$$

Proof. (i) Let $x \in S^+$, and let $\sigma_{x,\varepsilon}$ be a disk with the centre at x and radius ε sufficiently small so that $\bar{\sigma}_{x,\varepsilon}$ lies strictly inside S^+. By Green's second identity in $\bar{S}^+ \setminus \sigma_\varepsilon$ with $v(y) = g(x,y)$,

$$0 = \left(\int_{\partial S} + \int_{\partial \sigma_{x,\varepsilon}}\right)[g(x,y)(\partial_\nu u)(y) - (\partial_{\nu(y)}g)(x,y)u(y)]\,ds(y). \tag{1.9}$$

Using polar coordinates with the pole at x and noting that the outward normal on $\partial \sigma_{x,\varepsilon}$ is directed into $\sigma_{x,\varepsilon}$, we find that on $\partial \sigma_{x,\varepsilon}$

$$(\partial_{\nu(y)}g)(x,y) = -(2\pi)^{-1}\partial_{\nu(y)}\ln|x-y| = (2\pi)^{-1}(\partial_r \ln r)|_{r=\varepsilon} = (2\pi)^{-1}\varepsilon^{-1},$$

so the integral over $\partial \sigma_{x,\varepsilon}$ in (1.9) becomes

$$(2\pi)^{-1}\int_0^{2\pi}[(\ln\varepsilon)(\partial_r u)(\varepsilon,\theta) - \varepsilon^{-1}u(\varepsilon,\theta)]\varepsilon\,d\theta,$$

which tends to $-u(x)$ as $\varepsilon \to 0$. If $x \in \partial S$, then we need only the approximation of a half-disk to isolate x, and the limit is $-\tfrac{1}{2}u(x)$ (see the Appendix). If $x \in S^-$, then $\ln|x-y|$ is harmonic in S^+ and we do not need to make use of $\sigma_{x,\varepsilon}$.

(ii) For $x \in S^-$, consider a disk K_R with the centre at x and radius R sufficiently large so that $\bar{S}^+$ lies strictly inside K_R. By (i) applied to $\bar{S}^- \cap \bar{K}_R$,

$$\left(-\int_{\partial S} + \int_{\partial K_R}\right)\left[g(x,y)(\partial_\nu u)(y) - u(y)(\partial_{\nu(y)} g)(x,y)\right] ds(y) = u(x),$$

where we have taken account of the fact that for this domain the outward normal on ∂S is directed into S^+. Since $u \in \mathcal{A}$, using (1.5) and polar coordinates with the pole at x, we see that for R large the second integral on the left-hand side in the preceding equality is of the form

$$\int_0^{2\pi} \left[(\ln R) \cdot O(R^{-2}) + O(R^{-1}) \cdot R^{-1}\right] R \, d\theta,$$

which means that it vanishes as $R \to \infty$, yielding the required formula. Similar arguments are applied when $x \in \partial S$ or $x \in S^+$.

1.4. Uniqueness theorems

Before we solve the boundary value problems set out in §1.2, we need to know how many solutions we should expect in each case.

1.29. Theorem. (i) *Each of* (D^+), (R^+), (D^-), (N^-) *and* (R^-) *has at most one solution.*
(ii) *Any two solutions of* (N^+) *differ by a constant.*

Proof. The difference u of any two solutions satisfies the corresponding homogeneous boundary value problem. Thus, for (D^+)

$$(\Delta u)(x) = 0, \quad x \in S^+, \quad \text{and} \quad u(x) = 0, \quad x \in \partial S.$$

By Corollary 1.27(i), this leads to

$$\int_{S^+} |\operatorname{grad} u|^2 \, da = 0,$$

which, since $u \in C^1(\bar{S}^+)$, implies that $(\operatorname{grad} u)(x) = 0$, $x \in \bar{S}^+$; therefore, $u = \text{const}$ in $\bar{S}^+$, and the homogeneous boundary condition tells us that $u = 0$. The argument is similar for (D^-), where we make use of Corollary 1.27(ii).

In the case of (N$^+$), u still satisfies $(\operatorname{grad} u)(x) = 0$, $x \in \bar{S}^+$, but we end up with $u = \operatorname{const}$ in $\bar{S}^+$ because the boundary condition $(\partial_\nu u)(x) = 0$, $x \in \partial S$, offers no additional information. By contrast, in (N$^-$) the constant must be zero since $u \in \mathcal{A}$. The homogeneous boundary condition of (R$^+$) can be written as

$$(\partial_\nu u)(x) = -(\sigma u)(x), \quad x \in \partial S,$$

and from Corollary 1.27(i) it follows that

$$\int_{S^+} |\operatorname{grad} u|^2 \, da + \int_{\partial S} \sigma u^2 \, ds = 0;$$

since $\sigma > 0$, we conclude that $u(x) = 0$, $x \in \bar{S}^+$. The same is true in (R$^-$), where we operate with Corollary 1.27(ii) and the boundary condition

$$(\partial_\nu u)(x) = (\sigma u)(x), \quad x \in \partial S.$$

The following statement is almost obvious, but we mention it as a separate item because of its frequent use in the solution of the boundary value problems.

1.30. Corollary. (i) *If $u^\mathcal{A} + c$, $c = \operatorname{const}$, is a solution of the homogeneous problem* (D$^-$) *or* (R$^-$), *then $c = 0$ and $u^\mathcal{A} = 0$.*
(ii) *If $u \in \mathcal{A}$ is a solution of* (D$^-$) *with boundary data $u|_{\partial S} = c = \operatorname{const}$, then*

$$c = 0, \quad u = 0.$$

Proof. (i) By Theorem 1.29(i), the zero solution is the only solution of the homogeneous problem (D$^-$) or (R$^-$) in $\mathcal{A}^*$; hence, $u^\mathcal{A} + c = 0$, so $c = -u^\mathcal{A} = O(r^{-1})$. This implies that $c = 0$, which, in turn, yields $u^\mathcal{A} = 0$.

(ii) The assertion follows from (i) since $u - c$ is a solution of the homogeneous problem (D$^-$).

1.31. Remark. The existence of at most one solution can also be proved for (D$^\pm$) under the less stringent requirement that the solutions should be only $C(\bar{S}^\pm)$. The proof is based on the maximum principle (see, for example, [25]). Since we wanted to make use of Green's identities (by symmetry with the argument used for systems in Chapters 2 and 3, where a maximum principle is not available), we asked the solutions to be differentiable up to the boundary.

1.5. The harmonic potentials

Certain functions play a fundamental role in the solution of boundary value problems by means of boundary integral equation methods. In this section we briefly review these functions and their main properties. More details can be found in the Appendix.

1.32. Definition. The single-layer and double-layer potentials are defined, respectively, by

$$(V\varphi)(x) = \int_{\partial S} g(x, y)\varphi(y) \, ds(y), \tag{1.10}$$

$$(W\psi)(x) = \int_{\partial S} (\partial_{\nu(y)}g)(x, y)\psi(y) \, ds(y), \tag{1.11}$$

where φ and ψ are density functions with suitable smoothness properties.

1.33. Theorem. *Let $\varphi, \psi \in C(\partial S)$, and let p be the linear functional on $C(\partial S)$ defined by*

$$p\varphi = \int_{\partial S} \varphi \, ds.$$

Then
  (i) *$V\varphi \in \mathcal{A}$ if and only if $p\varphi = 0$;*
  (ii) *$W\psi \in \mathcal{A}$.*

Proof. For y fixed, as $|x| \to \infty$

$$\begin{aligned} \ln|x-y| &= \ln|x| + \sum_{n=1}^{\infty} \lambda_n(\theta, y) r^{-n}, \\ |x-y|^{-2} &= \sum_{n=1}^{\infty} \mu_n(\theta, y) r^{-n-1}, \end{aligned} \tag{1.12}$$

where (r, θ) are the polar coordinates of x and each of λ_n and μ_n is a finite linear combination of products of various powers of y_1 and y_2 with coefficients of the form

$$a_n + \sum_{l=1}^{k_n} (b_n^l \cos l\theta + c_n^l \sin l\theta), \quad a_n, b_n^l, c_n^l = \text{const}; \tag{1.13}$$

for example,

$$\lambda_1 = -(y_1 \cos\theta + y_2 \sin\theta), \quad \mu_1 = 1, \quad \text{etc.}$$

Noting that

$$\partial_{\nu(y)} \ln |x-y| = -\big[(x_1-y_1)\nu_1(y) + (x_2-y_2)\nu_2(y)\big]|x-y|^{-2}$$

and replacing (1.12) in (1.10) and (1.11), we find that

$$\begin{aligned}(V\varphi)(r,\theta) &= -(2\pi)^{-1}(p\varphi) + \sum_{n=1}^{\infty} \alpha_n(\theta) r^{-n},\\ (W\psi)(r,\theta) &= \sum_{n=1}^{\infty} \beta_n(\theta) r^{-n},\end{aligned} \qquad (1.14)$$

with the α_n and β_n of the form (1.13). The theorem now follows immediately from equalities (1.14).

1.34. Theorem. (i) *If $\varphi, \psi \in C(\partial S)$, then $(V\varphi)(x)$ and $(W\psi)(x)$ are analytic at all $x \in S^+ \cup S^-$ and*

$$\Delta(V\varphi)(x) = \Delta(W\psi)(x) = 0, \quad x \in S^+ \cup S^-.$$

(ii) *If $\varphi, \psi \in C^{0,\alpha}(\partial S)$, $\alpha \in (0,1)$, then the direct values $V_0 \varphi$ and $W_0 \psi$ of $V\varphi$ and $W\psi$ on ∂S exist (the latter as principal value). Also, the operators $\mathcal{V}^\pm$ defined by*

$$\mathcal{V}^+\varphi = (V\varphi)|_{\bar{S}^+}, \quad \mathcal{V}^-\varphi = (V\varphi)|_{\bar{S}^-}$$

map $C^{0,\alpha}(\partial S)$ to $C^{1,\alpha}(\bar{S}^\pm)$, $\alpha \in (0,1)$, respectively, and

$$\partial_\nu(\mathcal{V}^+\varphi) = \big(W_0^* + \tfrac{1}{2}I\big)\varphi, \quad \partial_\nu(\mathcal{V}^-\varphi) = \big(W_0^* - \tfrac{1}{2}I\big)\varphi, \quad \varphi \in C^{0,\alpha}(\partial S), \qquad (1.15)$$

where I is the identity operator and W_0^* is the adjoint of the direct value operator W_0, defined (in the sense of principal value) by

$$(W_0^*\varphi)(x) = \int_{\partial S} (\partial_{\nu(x)} g)(x,y)\varphi(y)\,ds(y), \quad x \in \partial S. \qquad (1.16)$$

(iii) *The operators $\mathcal{W}^\pm$ defined by*

$$\mathcal{W}^+\psi = \begin{cases} (W\psi)|_{S^+} & \text{in } S^+, \\ (W_0 - \tfrac{1}{2}I)\psi & \text{on } \partial S, \end{cases} \quad \mathcal{W}^-\psi = \begin{cases} (W\psi)|_{S^-} & \text{in } S^-, \\ (W_0 + \tfrac{1}{2}I)\psi & \text{on } \partial S \end{cases} \qquad (1.17)$$

map $C^{0,\alpha}(\partial S)$ to $\Gamma^{0,\alpha}(\bar{S}^\pm)$ and $C^{1,\alpha}(\partial S)$ to $C^{1,\alpha}(\bar{S}^\pm)$, $\alpha \in (0,1)$, respectively, and

$$\partial_\nu(\mathcal{W}^+\psi) = \partial_\nu(\mathcal{W}^-\psi), \quad \psi \in C^{1,\alpha}(\partial S). \qquad (1.18)$$

(iv) *The operator W_0 maps $C^{0,\alpha}(\partial S)$ to $C^{1,\alpha}(\partial S)$, $\alpha \in (0,1)$.*

The proof of these statements is based on Theorem 1.4 (see the Appendix), since

$$V\varphi = (2\pi)^{-1}(v\varphi), \quad W\psi = (2\pi)^{-1}(w\psi).$$

1.35. Remarks. (i) In Theorem 1.34, the derivatives on ∂S of functions defined in $\bar{S}^+$ or $\bar{S}^-$ are one-sided.

(ii) Obviously, $\Delta(\mathcal{V}^+\varphi)(x) = 0$, $x \in S^+$, and $\Delta(\mathcal{V}^-\varphi)(x) = 0$, $x \in S^-$.

(iii) If $\mathcal{V}^+\varphi = \mathcal{V}^-\varphi = 0$, then, by (1.15), $\varphi = 0$.

(iv) In view of Theorem 1.34(ii),(iii), Green's representation formulae for a harmonic function u in S^+ and a harmonic function $u \in \mathcal{A}$ in S^-, respectively, can now be written as

$$\begin{aligned} \mathcal{V}^+(\partial_\nu u) - \mathcal{W}^+(u|_{\partial S}) &= u, \\ \mathcal{V}^-(\partial_\nu u) - \mathcal{W}^-(u|_{\partial S}) &= 0 \end{aligned} \quad (1.19)$$

and

$$\begin{aligned} -\mathcal{V}^-(\partial_\nu u) + \mathcal{W}^-(u|_{\partial S}) &= u, \\ -\mathcal{V}^+(\partial_\nu u) + \mathcal{W}^+(u|_{\partial S}) &= 0. \end{aligned} \quad (1.20)$$

(v) From (iv) with $u = 1$ it follows that

$$\mathcal{W}^+ 1 = -1, \quad W_0 1 = -\tfrac{1}{2}, \quad \mathcal{W}^- 1 = 0.$$

(vi) Equality (1.18) enables us to define a boundary operator

$$N_0 : C^{1,\alpha}(\partial S) \to C^{0,\alpha}(\partial S)$$

by setting

$$N_0 \psi = \partial_\nu(\mathcal{W}^+ \psi) = \partial_\nu(\mathcal{W}^- \psi), \quad \psi \in C^{1,\alpha}(\partial S). \quad (1.21)$$

In accordance with Theorem 1.34(ii),(iii), in what follows we use the notation

$$(\mathcal{V}^\pm \varphi)|_{\partial S} = \mathcal{V}_0^\pm \varphi = V_0 \varphi, \quad (\mathcal{W}^\pm \psi)|_{\partial S} = \mathcal{W}_0^\pm \psi = (W_0 \mp \tfrac{1}{2} I)\psi. \quad (1.22)$$

From now on in this chapter we assume that the boundary integral operators V_0, W_0, W_0^*, $W_0 \pm \tfrac{1}{2} I$ and $W_0^* \pm \tfrac{1}{2} I$ are defined on $C^{0,\alpha}(\partial S)$, while N_0 is defined on $C^{1,\alpha}(\partial S)$, $\alpha \in (0,1)$. Any other arrangement will be mentioned explicitly. To avoid cumbersome notation, we also write, for example, $\Delta \mathcal{U}^+ = 0$ instead of the longer version $(\Delta \mathcal{U}^+)(x) = 0$, $x \in S^+$, when $\mathcal{U}^+$ is a function defined in $\bar{S}^+$. The same is done in the case of functions defined on $\bar{S}^-$, ∂S and $\mathbb{R}^2$.

1.6. Properties of the boundary operators

The boundary operators introduced in Theorem 1.34(ii),(iii) and Remark 1.35(vi) have a number of very important properties, which we use extensively in the solution of the boundary value problems.

1.36. Theorem. V_0, W_0, W_0^* and N_0 satisfy the composition formulae

$$W_0 V_0 = V_0 W_0^*, \qquad N_0 V_0 = W_0^{*2} - \tfrac{1}{4}I \quad on \ C^{0,\alpha}(\partial S), \tag{1.23}$$

$$N_0 W_0 = W_0^* N_0, \qquad V_0 N_0 = W_0^2 - \tfrac{1}{4}I \quad on \ C^{1,\alpha}(\partial S). \tag{1.24}$$

Proof. Let φ be arbitrary in $C^{0,\alpha}(\partial S)$. By Theorem 1.34(ii), we can write

$$\begin{aligned} \mathcal{V}_0^+ \varphi &= V_0 \varphi = \beta \in C^{1,\alpha}(\partial S), \\ \partial_\nu (\mathcal{V}^+ \varphi) &= \gamma \in C^{0,\alpha}(\partial S). \end{aligned} \tag{1.25}$$

Since $\mathcal{V}^+ \varphi$ is harmonic in S^+, it admits the representation $(1.19)_1$, which, by (1.25), becomes

$$\mathcal{V}^+ \varphi = \mathcal{V}^+ \gamma - \mathcal{W}^+ \beta. \tag{1.26}$$

All the terms in (1.26) belong to $C^{1,\alpha}(\bar{S}^+)$, so we can take their normal derivatives on ∂S to obtain (in view of (1.25), (1.15) and (1.21))

$$\gamma = (W_0^* + \tfrac{1}{2}I)\gamma - N_0 \beta,$$

or

$$N_0 \beta = (W_0^* - \tfrac{1}{2}I)\gamma. \tag{1.27}$$

Similarly, by (1.17), restricting (1.26) to ∂S yields

$$\beta = V_0 \gamma - (W_0 - \tfrac{1}{2}I)\beta,$$

or

$$(W_0 + \tfrac{1}{2}I)\beta = V_0 \gamma. \tag{1.28}$$

On the other hand, using $(1.15)_1$, we can write $(1.25)_2$ as

$$\gamma = (W_0^* + \tfrac{1}{2}I)\varphi. \tag{1.29}$$

Replacing β and γ from $(1.25)_1$ and (1.29) in (1.27) and (1.28), we find that

$$\begin{aligned} N_0(V_0 \varphi) &= (W_0^* - \tfrac{1}{2}I)(W_0^* + \tfrac{1}{2}I)\varphi, \\ (W_0 + \tfrac{1}{2}I)(V_0 \varphi) &= V_0(W_0^* + \tfrac{1}{2}I)\varphi, \end{aligned}$$

from which the arbitrariness of φ in $C^{0,\alpha}(\partial S)$ produces (1.23).

Equalities (1.24) are derived similarly, the argument being based on the harmonic function $\mathcal{W}^+\psi$ with ψ arbitrary in $C^{1,\alpha}(\partial S)$.

1.37. Theorem. *If there is a function $\varphi \in C^{0,\alpha}(\partial S)$, $\varphi \neq 0$, such that*

$$(W_0^* + \tfrac{1}{2}I)\varphi = 0,$$

then

$$\mathcal{V}^+\varphi = c = \mathrm{const}, \quad p\varphi \neq 0. \tag{1.30}$$

Proof. By $(1.15)_1$, $(W_0^* + \tfrac{1}{2}I)\varphi = 0$ implies that $\partial_\nu(\mathcal{V}^+\varphi) = 0$. Hence, $\mathcal{V}^+\varphi$ is a solution of the homogeneous interior Neumann problem, so from Theorem 1.29(ii) we obtain $(1.30)_1$.

Suppose now that $p\varphi = 0$. Then, by Theorems 1.33(i) and 1.34(i),(ii), $\mathcal{V}^-\varphi$ is a solution of the exterior Dirichlet problem

$$\Delta(\mathcal{V}^-\varphi) = 0,$$
$$\mathcal{V}_0^-\varphi = \mathcal{V}_0^+\varphi = c,$$
$$\mathcal{V}^-\varphi \in \mathcal{A};$$

hence, by Corollary 1.30(ii), $\mathcal{V}^-\varphi = c = 0$. Since we have also seen in $(1.30)_1$ that $\mathcal{V}^+\varphi = c = 0$, we use Remark 1.35(iii) to deduce that $\varphi = 0$, which contradicts our assumption; consequently, $p\varphi \neq 0$.

1.38. Theorem. (i) *The null spaces of $W_0 - \tfrac{1}{2}I$ and $W_0^* - \tfrac{1}{2}I$ consist of the zero function.*

(ii) *The null spaces of $W_0 + \tfrac{1}{2}I$ and $W_0^* + \tfrac{1}{2}I$ are one-dimensional; they are spanned, respectively, by 1 and a function $\Phi \in C^{0,\alpha}(\partial S)$ such that $p\Phi = 1$.*

Proof. (i) Let ψ_0 be such that $(W_0^* - \tfrac{1}{2}I)\psi_0 = 0$, that is,

$$\int_{\partial S} (\partial_{\nu(x)}g)(x,y)\psi_0(y)\,ds(y) - \tfrac{1}{2}\psi_0(x) = 0, \quad x \in \partial S.$$

Multiplying this equation by $ds(x)$, integrating it term by term with respect to x over ∂S, changing the order of integration and taking Remark 1.35(v) into account, we find that

$$0 = \int_{\partial S} \left(\int_{\partial S} (\partial_{\nu(x)} g)(x,y)\, ds(x) \right) \psi_0(y)\, ds(y) - \tfrac{1}{2} \int_{\partial S} \psi_0(x)\, ds(x)$$
$$= \int_{\partial S} (W_0 1) \psi_0\, ds - \tfrac{1}{2} \int_{\partial S} \psi_0\, ds = - \int_{\partial S} \psi_0\, ds = -p\psi_0.$$

Then, by Theorem 1.33(i), $\mathcal{V}^- \psi_0 \in \mathcal{A}$. Also, by Theorem 1.34(i),(iii), $\Delta(\mathcal{V}^- \psi_0) = 0$ in S^- and $(W_0^* - \tfrac{1}{2} I)\psi_0 = 0$ is equivalent to

$$\partial_\nu (\mathcal{V}^- \psi_0) = 0,$$

so $\mathcal{V}^- \psi_0$ is a solution of the homogeneous problem (N$^-$). Since, according to Theorem 1.29(i), this problem has at most one solution, we conclude that $\mathcal{V}^- \psi_0 = 0$.

Next, by (1.22)$_1$,

$$\mathcal{V}_0^+ \psi_0 = \mathcal{V}_0 \psi_0 = \mathcal{V}_0^- \psi_0 = 0.$$

In addition, $\Delta(\mathcal{V}^+ \psi_0) = 0$; hence, $\mathcal{V}^+ \psi_0$ is a solution of the homogeneous problem (D$^+$). Again by Theorem 1.29(i), $\mathcal{V}^+ \psi_0 = 0$, which, in view of Remark 1.35(iii), means that $\psi_0 = 0$; therefore, the null space of $W_0^* - \tfrac{1}{2} I$ consists only of the zero function. By Theorem 1.14 and Remark 1.8, we can apply the Fredholm Alternative and deduce that the null space of $W_0 - \tfrac{1}{2} I$ has the same property.

(ii) By Remark 1.35(v),

$$(W_0 + \tfrac{1}{2} I) a = 0 \quad \text{for all } a = \text{const}.$$

This implies that the dimension of the null space of $W_0 + \tfrac{1}{2} I$ is at least 1. By the Fredholm Alternative, the same is true for $W_0^* + \tfrac{1}{2} I$; therefore, there is a nonzero $\varphi_0 \in C^{0,\alpha}(\partial S)$ such that $(W_0^* + \tfrac{1}{2} I)\varphi_0 = 0$. By Theorem 1.37, φ_0 satisfies $\mathcal{V}^+ \varphi_0 = c = \text{const}$ and $p\varphi_0 \neq 0$.

Let $\tilde{\varphi}_0$ be another function in the null space of $W_0^* + \tfrac{1}{2} I$. By the same argument as above, $\tilde{\varphi}_0$ satisfies $\mathcal{V}^+ \tilde{\varphi}_0 = \tilde{c} = \text{const}$ and $p\tilde{\varphi}_0 \neq 0$. Since

$$p\big((p\tilde{\varphi}_0)\varphi_0 - (p\varphi_0)\tilde{\varphi}_0\big) = (p\tilde{\varphi}_0)(p\varphi_0) - (p\varphi_0)(p\tilde{\varphi}_0) = 0,$$

we have

$$\mathcal{V}^+\big((p\tilde{\varphi}_0)\varphi_0 - (p\varphi_0)\tilde{\varphi}_0\big) = (p\tilde{\varphi}_0)c - (p\varphi_0)\tilde{c}, \tag{1.31}$$

while $\mathcal{V}^-\big((p\tilde{\varphi}_0)\varphi_0 - (p\varphi_0)\tilde{\varphi}_0\big)$ satisfies

$$(\Delta \mathcal{V}^-)\big((p\tilde{\varphi}_0)\varphi_0 - (p\varphi_0)\tilde{\varphi}_0\big) = 0,$$

$$\mathcal{V}_0^-\big((p\tilde{\varphi}_0)\varphi_0 - (p\varphi_0)\tilde{\varphi}_0\big) = \mathcal{V}_0^+\big((p\tilde{\varphi}_0)\varphi_0 - (p\varphi_0)\tilde{\varphi}_0\big) = (p\tilde{\varphi}_0)c - (p\varphi_0)\tilde{c},$$
$$\mathcal{V}^-\big((p\tilde{\varphi}_0)\varphi_0 - (p\varphi_0)\tilde{\varphi}_0\big) \in \mathcal{A}.$$

By Corollary 1.30(ii),
$$\mathcal{V}^-\big((p\tilde{\varphi}_0)\varphi_0 - (p\varphi_0)\tilde{\varphi}_0\big) = (p\tilde{\varphi}_0)c - (p\varphi_0)\tilde{c} = 0.$$

From this equality, (1.31), and Remark 1.35(iii), it follows that
$$(p\tilde{\varphi}_0)\varphi_0 - (p\varphi_0)\tilde{\varphi}_0 = 0.$$

Since $p\tilde{\varphi}_0 \neq 0$, we can write
$$\tilde{\varphi}_0 = (p\varphi_0)(p\tilde{\varphi}_0)^{-1}\varphi_0,$$

which means that the null space of $W_0^* + \frac{1}{2}I$ is one-dimensional and spanned by φ_0. Hence, the null space of $W_0 + \frac{1}{2}I$ is also one-dimensional and spanned by 1.

For convenience, without loss of generality, we choose φ_0 to be the unique function Φ in the null space of $W_0^* + \frac{1}{2}I$ which satisfies
$$(1, \Phi) = p\Phi = 1.$$

1.39. Theorem. (i) *The null spaces of $W_0^2 - \frac{1}{4}I$ and $W_0^{*2} - \frac{1}{4}I$ coincide with those of $W_0 + \frac{1}{2}I$ and $W_0^* + \frac{1}{2}I$ (that is, they are spanned by 1 and Φ, respectively).*
(ii) *$N_0\psi = 0$ if and only if $\psi = c = \text{const}$.*

Proof. (i) The result follows from the fact that, by Theorem 1.38(i),
$$(W_0^2 - \tfrac{1}{4}I)\varphi = (W_0 - \tfrac{1}{2}I)\big((W_0 + \tfrac{1}{2}I)\varphi\big) = 0$$

implies that $(W_0 + \frac{1}{2}I)\varphi = 0$, which, in turn, implies that $\varphi = \text{const}$.

The second part of the statement is proved by a symmetric argument.
(ii) If $N_0\psi = 0$, then, by $(1.24)_2$,
$$0 = V_0(N_0\psi) = (V_0 N_0)\psi = (W_0^2 - \tfrac{1}{4}I)\psi,$$

and (i) yields $\psi = \text{const}$. Conversely, by (1.21) and Remark 1.35(v), for any $c \in \mathbb{R}$
$$N_0 c = c(N_0 1) = c\partial_\nu(\mathcal{W}^+ 1) = c\partial_\nu(-1) = 0.$$

In the proof of Theorem 1.38(ii) we have shown the existence of a non-zero function $\Phi \in C^{0,\alpha}(\partial S)$ such that $V_0\Phi = \text{const}$ and $p\Phi = 1$. This assertion can be sharpened and used to define a certain characteristic of the boundary curve.

PROPERTIES OF THE BOUNDARY OPERATORS

1.40. Theorem. *For every simple closed C^2-curve ∂S and any $\alpha \in (0,1)$, there are a unique non-zero function $\Phi \in C^{0,\alpha}(\partial S)$ and a unique constant ω such that*

$$V_0 \Phi = \omega, \quad p\Phi = 1. \tag{1.32}$$

Proof. As already mentioned, we have seen above that such a pair exists. To verify that it is unique, let $\tilde{\Phi}$, $\tilde{\omega}$ be another pair with the same properties. Since $\mathcal{V}^+\Phi$ and $\mathcal{V}^+\tilde{\Phi}$ are solutions of the interior Dirichlet problems with (constant) boundary data ω and $\tilde{\omega}$, respectively, from Theorem 1.29(i) it follows that

$$\mathcal{V}^+\Phi = \omega, \quad \mathcal{V}^+\tilde{\Phi} = \tilde{\omega};$$

hence,

$$\mathcal{V}^+(\Phi - \tilde{\Phi}) = \omega - \tilde{\omega}.$$

Then, as $p(\Phi - \tilde{\Phi}) = 1 - 1 = 0$ and

$$\mathcal{V}_0^-(\Phi - \tilde{\Phi}) = \mathcal{V}_0^+(\Phi - \tilde{\Phi}) = \omega - \tilde{\omega},$$

the function $\mathcal{V}^-(\Phi - \tilde{\Phi}) \in \mathcal{A}$ is the (unique) solution of the exterior Dirichlet problem with boundary data $\omega - \tilde{\omega} = \text{const}$; therefore, by Corollary 1.30(ii),

$$\mathcal{V}^-(\Phi - \tilde{\Phi}) = 0, \quad \omega - \tilde{\omega} = 0,$$

which means that $\omega = \tilde{\omega}$. This, in turn, yields

$$\mathcal{V}^+(\Phi - \tilde{\Phi}) = \mathcal{V}^-(\Phi - \tilde{\Phi}) = 0;$$

so, by Remark 1.35(iii), $\Phi - \tilde{\Phi} = 0$, as required.

1.41. Corollary. *If $V_0\varphi = c = \text{const}$, then $c = k\omega$ and $\varphi = k\Phi$ for some constant k.*

Proof. By $(1.23)_2$ and Theorem 1.39(ii),

$$(N_0 V_0)\varphi = N_0(V_0\varphi) = N_0 c = 0 = (W_0^{*2} - \tfrac{1}{4}I)\varphi,$$

and from Theorems 1.39(i) and 1.38(ii) it follows that $\varphi = k\Phi$, $k = \text{const}$. Then, in view of $(1.32)_1$,

$$c = V_0\varphi = k(V_0\Phi) = k\omega.$$

1.42. Remarks. (i) Because of the uniqueness of the solution of (D^+) and the fact that $\Delta(\mathcal{V}^+\Phi) = \Delta\omega = 0$ (in S^+) and $\mathcal{V}_0^+\Phi = V_0\Phi = \omega$, we have $\mathcal{V}^+\Phi = \omega$.

(ii) The numbers $2\pi\omega$ and $e^{-2\pi\omega}$ are called Robin's constant and the logarithmic capacity of ∂S, respectively [17]. For example, for a circle ∂K_R of radius R [18]

$$V_0 1 = -(2\pi)^{-1} \int_{\partial K_R} \ln|x-y|\, ds(y) = -R\ln R \quad \text{for all } x \in \partial S,$$

$$p1 = \int_{\partial S} 1\, ds = \int_0^{2\pi} R\, d\theta = 2\pi R,$$

so

$$\Phi = (2\pi R)^{-1}, \quad \omega = -(2\pi)^{-1}\ln R$$

and the logarithmic capacity is R.

This means that if the logarithmic capacity of ∂S is 1, then $\omega = 0$ and the null space of the boundary operator V_0 contains non-zero functions. In fact, we can give a full description of this space.

1.43. Theorem. *If the logarithmic capacity of ∂S is 1 (that is, $\omega = 0$), then the null space of V_0 coincides with that of $W_0^* + \tfrac{1}{2}I$. In all other cases, this space consists of zero alone.*

Proof. Suppose that $\omega = 0$. By Theorem 1.37, any function φ in the null space of $W_0^* + \tfrac{1}{2}I$ satisfies $V_0\varphi = c = \text{const}$. By Corollary 1.41, this implies that $c = k\omega = 0$, so φ belongs to the null space of V_0. Conversely, if $V_0\varphi = 0$, then, using the composition formula $(1.23)_2$, we find that

$$0 = N_0(V_0\varphi) = (N_0 V_0)\varphi = (W_0^{*2} - \tfrac{1}{4}I)\varphi,$$

which, by Theorem 1.39(i), implies that φ belongs to the null space of $W_0^* + \tfrac{1}{2}I$.

Suppose now that $\omega \neq 0$ and that $V_0\varphi = 0$, and consider the function

$$\mathcal{U}^- = \mathcal{V}^-\big(\varphi - (p\varphi)\Phi\big) + \omega(p\varphi).$$

Since

$$p\big(\varphi - (p\varphi)\Phi\big) = p\varphi - p\varphi = 0,$$

we have the asymptotic formula

$$\mathcal{U}^-(x) = \omega(p\varphi) + \mathcal{U}^{\mathcal{A}}(x) \quad \text{as } |x| \to \infty.$$

Also, since $V_0\varphi = 0$,

$$\mathcal{U}^-(x) = \big(V_0\varphi - (p\varphi)(V_0\Phi) + \omega(p\varphi)\big)(x) = -\omega(p\varphi) + \omega(p\varphi) = 0, \quad x \in \partial S,$$

and $\mathcal{U}^-$ is harmonic in S^-. By Corollary 1.30(i),

$$\omega(p\varphi) = 0, \quad \mathcal{U}^- = \mathcal{V}^-\big(\varphi - (p\varphi)\Phi\big) = 0.$$

In turn, this implies that $p\varphi = 0$ and $\mathcal{V}^-\varphi = 0$. Then, since $\mathcal{V}_0^+\varphi = \mathcal{V}_0^-\varphi = 0$, we deduce that $\mathcal{V}^+\varphi = 0$ as the unique solution of the homogeneous problem (D$^+$). Remark 1.35(iii) now yields $\varphi = 0$.

1.44. Lemma. (i) *If q is the functional defined on $C(\partial S)$ by*

$$q\varphi = \int_{\partial S} \Phi\varphi\, ds, \qquad (1.33)$$

then for any $\varphi \in C(\partial S)$

$$q(V_0\varphi) = \omega(p\varphi).$$

(ii) *For any $\psi \in C^{0,\alpha}(\partial S)$*

$$p(W_0^*\psi) = W_0(p\psi) = -\tfrac{1}{2}p\psi.$$

(iii) *For any $\psi \in C^{0,\alpha}(\partial S)$*

$$q(W_0\psi) = -\tfrac{1}{2}q\psi.$$

(iv) *For any $\psi \in C^{1,\alpha}(\partial S)$*

$$p(N_0\psi) = 0.$$

Proof. (i) Changing the order of integration, we see that

$$q(V_0\varphi) = \int_{\partial S} \Phi(V_0\varphi)\, ds = \int_{\partial S} \Phi(x)\left(\int_{\partial S} g(x,y)\varphi(y)\, ds(y)\right) ds(x)$$

$$= \int_{\partial S}\left(\int_{\partial S} g(x,y)\Phi(x)\, ds(x)\right)\varphi(y)\, ds(y)$$

$$= \int_{\partial S} (V_0\Phi)\varphi\, ds = \omega\int_{\partial S} \varphi\, ds = \omega(p\varphi).$$

(ii) By (1.16) and Remark 1.35(v),

$$\begin{aligned}
p(W_0^*\psi) &= \int_{\partial S}\left(\int_{\partial S}(\partial_{\nu(x)}g)(x,y)\psi(y)\,ds(y)\right)ds(x) \\
&= \int_{\partial S}\left(\int_{\partial S}(\partial_{\nu(x)}g)(y,x)\,ds(x)\right)\psi(y)\,ds(y) \\
&= \int_{\partial S}(W_0 1)\psi\,ds = (p\psi)(W_0 1) = W_0(p\psi) = -\tfrac{1}{2}p\psi.
\end{aligned}$$

(iii) Changing the order of integration, we obtain

$$\begin{aligned}
q(W_0\psi) &= \int_{\partial S}\Phi(x)\left(\int_{\partial S}(\partial_{\nu(y)}g)(x,y)\psi(y)\,ds(y)\right)ds(x) \\
&= \int_{\partial S}\left(\int_{\partial S}(\partial_{\nu(y)}g)(x,y)\Phi(x)\,ds(x)\right)\psi(y)\,ds(y) = \int_{\partial S}(W_0^*\Phi)\psi\,ds.
\end{aligned}$$

Since, by Theorem 1.38(ii), $W_0^*\Phi = -\tfrac{1}{2}\Phi$, we can write

$$q(W_0\psi) = -\tfrac{1}{2}\int_{\partial S}\Phi\psi\,ds = -\tfrac{1}{2}q\psi.$$

(iv) By the divergence theorem,

$$p(N_0\psi) = p(\partial_\nu(\mathcal{W}^+\psi)) = \int_{\partial S}\partial_\nu(\mathcal{W}^+\psi)\,ds = \int_{S^+}\Delta(\mathcal{W}^+\psi)\,da = 0.$$

1.7. The classical indirect method

First we discuss the interior and exterior Dirichlet and Neumann problems. We seek the solutions of these problems in the form

$$\begin{aligned}
u &= \mathcal{W}^+\varphi && \text{for } (\text{D}^+), \\
u &= \mathcal{V}^+\psi && \text{for } (\text{N}^+), \\
u &= \mathcal{W}^-\varphi + c && \text{for } (\text{D}^-), \\
u &= \mathcal{V}^-\psi && \text{for } (\text{N}^-),
\end{aligned}$$

where $c = \text{const}$.

From the properties of $\mathcal{V}^\pm$ and $\mathcal{W}^\pm$ in Theorem 1.34 and the boundary conditions in each of these problems, it follows that the unknown densities φ and ψ must satisfy the boundary integral equations

$$(W_0 - \tfrac{1}{2}I)\varphi = \mathcal{P} \quad \text{for (D}^+\text{)}, \qquad (\mathcal{D}_C^+)$$

$$(W_0^* + \tfrac{1}{2}I)\psi = \mathcal{Q} \quad \text{for (N}^+\text{)}, \qquad (\mathcal{N}_C^+)$$

$$(W_0 + \tfrac{1}{2}I)\varphi = \mathcal{R} - c \quad \text{for (D}^-\text{)}, \qquad (\mathcal{D}_C^-)$$

$$(W_0^* - \tfrac{1}{2}I)\psi = \mathcal{S} \quad \text{for (N}^-\text{)}. \qquad (\mathcal{N}_C^-)$$

It is obvious that $(\mathcal{D}_C^+), (\mathcal{N}_C^-)$ and $(\mathcal{D}_C^-), (\mathcal{N}_C^+)$ are mutually adjoint. Since the kernels of these Fredholm integral equations of the second kind are weakly singular (see Remark 1.8), the corresponding integral operators are compact and the Fredholm Alternative can be applied in the (real) dual system $\bigl(C^{0,\alpha}(\partial S), C^{0,\alpha}(\partial S)\bigr)$, $\alpha \in (0,1)$, with the bilinear form

$$(\varphi, \psi) = \int_{\partial S} \varphi \psi \, ds.$$

1.45. Theorem. (i) $(\mathcal{D}_C^+)$ has a unique solution $\varphi \in C^{1,\alpha}(\partial S)$ for any prescribed $\mathcal{P} \in C^{1,\alpha}(\partial S)$.

Then (D$^+$) has the (unique) solution

$$u = \mathcal{W}^+ \varphi.$$

(ii) $(\mathcal{D}_C^-)$ with $c = q\mathcal{R}$ is solvable in $C^{1,\alpha}(\partial S)$ for any $\mathcal{R} \in C^{1,\alpha}(\partial S)$, and its solution is unique up to an arbitrary constant.

Then (D$^-$) has the (unique) solution (in $\mathcal{A}^*$)

$$u = \mathcal{W}^- \varphi + q\mathcal{R},$$

where φ is any solution of $(\mathcal{D}_C^-)$.

(iii) $(\mathcal{N}_C^+)$ is solvable in $C^{0,\alpha}(\partial S)$ for any $\mathcal{Q} \in C^{0,\alpha}(\partial S)$ such that $p\mathcal{Q} = 0$. In this case the solution is unique up to a term of the form $a\Phi$, where a is an arbitrary constant.

Then (N$^+$) has the family of solutions

$$u = \mathcal{V}^+ \psi + c,$$

where ψ is any solution of $(\mathcal{N}_C^+)$ and c is an arbitrary constant.

(iv) ($\mathcal{N}_C^-$) has a unique solution $\psi \in C^{0,\alpha}(\partial S)$ for any $\mathcal{S} \in C^{0,\alpha}(\partial S)$.
Then, if $p\mathcal{S} = 0$, (N$^-$) has the (unique) solution (in $\mathcal{A}$)

$$u = \mathcal{V}^-\psi.$$

Proof. (i), (iv) By Theorem 1.38(i), the null spaces of $W_0 - \frac{1}{2}I$ and $W_0^* - \frac{1}{2}I$ consist of zero alone, so, by the Fredholm Alternative, ($\mathcal{D}_C^+$) and ($\mathcal{N}_C^-$) have unique solutions $\varphi, \psi \in C^{0,\alpha}(\partial S)$.

In the case of ($\mathcal{D}_C^+$), since $\mathcal{P} \in C^{1,\alpha}(\partial S)$, from Theorem 1.34(iv) it follows that $W_0\varphi \in C^{1,\alpha}(\partial S)$. Then $\varphi = 2(W_0\varphi + \mathcal{P}) \in C^{1,\alpha}(\partial S)$, which, in turn, means that $\mathcal{W}^+\varphi \in C^{1,\alpha}(\bar{S}^+)$. Consequently, by Definition 1.21 and Theorem 1.29(i), $u = \mathcal{W}^+\varphi$ is the (unique) solution of (D$^+$).

In the case of ($\mathcal{N}_C^-$), we apply p to both sides of the equation and make use of Lemma 1.44(ii) to find that

$$p\psi = -p\mathcal{S}.$$

Hence, if $p\mathcal{S} = 0$, then $p\psi = 0$, so $\mathcal{V}^-\psi \in \mathcal{A}$. Since $\Delta(\mathcal{V}^-\psi) = 0$ and, by Theorem 1.34(ii), $\mathcal{V}^-\psi \in C^{1,\alpha}(\bar{S}^-)$, Definition 1.21 shows that $u = \mathcal{V}^-\psi$ is the (unique) solution of (N$^-$).

(iii) Since the null space of $W_0 + \frac{1}{2}I$ is spanned by 1 (Theorem 1.38(ii)) and

$$(1, \mathcal{Q}) = \int_{\partial S} \mathcal{Q}\, ds = p\mathcal{Q} = 0,$$

we conclude that, by the Fredholm Alternative, ($\mathcal{N}_C^+$) is solvable in $C^{0,\alpha}(\partial S)$. The null space of $W_0^* + \frac{1}{2}I$ is spanned by Φ (Theorem 1.38(ii)), so the solution of ($\mathcal{N}_C^+$) is unique up to a term of the form $a\Phi$, where a is an arbitrary constant.

For any solution $\psi \in C^{0,\alpha}(\partial S)$ of ($\mathcal{N}_C^+$) and any $c = \text{const}$ we have $\Delta(\mathcal{V}^+\psi + c) = 0$ and $\mathcal{V}^+\psi + c \in C^{1,\alpha}(\bar{S}^+)$, so

$$u = \mathcal{V}^+\psi + c \tag{1.34}$$

is a solution of (N$^+$). By Remark 1.42(i),

$$\mathcal{V}^+(a\Phi) = a(\mathcal{V}^+\Phi) = a\omega = \text{const},$$

which is absorbed into the second (arbitrary) term on the right-hand side in (1.34).

(ii) If we take $c = \int_{\partial S} \Phi \mathcal{R}\, ds = q\mathcal{R}$, then

$$(\Phi, \mathcal{R} - c) = \int_{\partial S} \Phi(\mathcal{R} - c)\, ds = q\mathcal{R} - c(p\Phi) = c - c = 0,$$

so, by the Fredholm Alternative, $(\mathcal{D}_C^-)$ is solvable in $C^{0,\alpha}(\partial S)$ and, since the null space of $W_0 + \tfrac{1}{2}I$ is spanned by 1, its solution is unique up to a constant term a. As in the case of $(\mathcal{D}_C^+)$, φ in fact belongs to $C^{1,\alpha}(\partial S)$, which means that, by Theorem 1.34(iii), $\mathcal{W}^-\varphi \in C^{1,\alpha}(\bar{S}^-)$. Also, $\Delta(\mathcal{W}^-\varphi + q\mathcal{R}) = 0$ and, by Theorem 1.33(ii), $\mathcal{W}^-\varphi + q\mathcal{R} \in \mathcal{A}^*$; hence, $u = \mathcal{W}^-\varphi + q\mathcal{R}$ is the (unique) solution of (D^-). The arbitrary term a in φ does not affect u since, by Remark 1.35(v), $\mathcal{W}^-a = 0$.

We now turn our attention to the Robin problems. We seek the solution of (R^+) in the form
$$u = \mathcal{V}^+\big(\varphi - (p\varphi)\Phi\big) + p\varphi. \tag{1.35}$$
Then the boundary condition $(\partial_\nu u + \sigma u)(x) = \mathcal{K}(x)$, $x \in \partial S$, yields the boundary integral equation
$$(W_0^* + \tfrac{1}{2}I)\big(\varphi - (p\varphi)\Phi\big) + \sigma V_0\big(\varphi - (p\varphi)\Phi\big) + \sigma(p\varphi) = \mathcal{K}. \tag{$\mathcal{R}_C^+$}$$
Similarly, seeking the solution of (R^-) as
$$u = \mathcal{V}^-\big(\varphi - (p\varphi)\Phi\big) + p\varphi, \tag{1.36}$$
we use the condition $(\partial_\nu u - \sigma u)(x) = \mathcal{L}(x)$, $x \in \partial S$, to arrive at the boundary equation
$$(W_0^* - \tfrac{1}{2}I)\big(\varphi - (p\varphi)\Phi\big) - \sigma V_0\big(\varphi - (p\varphi)\Phi\big) - \sigma(p\varphi) = \mathcal{L}. \tag{$\mathcal{R}_C^-$}$$
Both $(\mathcal{R}_C^+)$ and $(\mathcal{R}_C^-)$ are Fredholm equations of the second kind.

1.46. Theorem. *Suppose that $\sigma \in C^{0,\alpha}(\partial S)$, $\alpha \in (0,1)$.*
(i) *$(\mathcal{R}_C^+)$ has a unique solution $\varphi \in C^{0,\alpha}(\partial S)$ for any $\mathcal{K} \in C^{0,\alpha}(\partial S)$. Then the (unique) solution of (R^+) is given by (1.35).*
(ii) *$(\mathcal{R}_C^-)$ has a unique solution $\varphi \in C^{0,\alpha}(\partial S)$ for any $\mathcal{L} \in C^{0,\alpha}(\partial S)$. Then the (unique) solution of (R^-) is given by (1.36).*

Proof. The operators on the left-hand side of both $(\mathcal{R}_C^+)$ and $(\mathcal{R}_C^-)$ map $C^{0,\alpha}(\partial S)$ to $C^{0,\alpha}(\partial S)$.

(i) Let φ_0 be a solution of the homogeneous equation $(\mathcal{R}_C^+)$, that is,
$$(W_0^* + \tfrac{1}{2}I)\big(\varphi_0 - (p\varphi_0)\Phi\big) + \sigma V_0\big(\varphi_0 - (p\varphi_0)\Phi\big) + \sigma(p\varphi_0) = 0.$$
Then $\mathcal{V}^+\big(\varphi_0 - (p\varphi_0)\Phi\big) + p\varphi_0$ is a solution of the homogeneous problem (R^+), so
$$\mathcal{V}^+\big(\varphi_0 - (p\varphi_0)\Phi\big) + p\varphi_0 = 0. \tag{1.37}$$

On the other hand, since, by $(1.32)_2$, $p(\varphi_0 - (p\varphi_0)\Phi) = 0$, from (1.22) and (1.37) it follows that the function

$$\mathcal{U}^- = \mathcal{V}^-(\varphi_0 - (p\varphi_0)\Phi) + p\varphi_0$$

satisfies

$$\Delta \mathcal{U}^- = 0,$$

$$\mathcal{U}^-|_{\partial S} = V_0(\varphi_0 - (p\varphi_0)\Phi) + p\varphi_0 = V_0^+(\varphi_0 - (p\varphi_0)\Phi) + p\varphi_0 = 0,$$
$$\mathcal{U}^-(x) = p(\varphi_0 - (p\varphi_0)\Phi)\ln|x| + p\varphi_0 + \mathcal{U}^{\mathcal{A}}(x) = \mathcal{U}^{\mathcal{A}}(x) + p\varphi_0 \quad \text{as } |x| \to \infty.$$

By Corollary 1.30(i), we then have

$$p\varphi_0 = 0, \quad \mathcal{V}^-(\varphi_0 - (p\varphi_0)\Phi) = \mathcal{V}^-\varphi_0 = 0;$$

also, from (1.37) it follows that $\mathcal{V}^+\varphi_0 = 0$, and Remark 1.35(iii) now yields $\varphi_0 = 0$.

By the Fredholm Alternative, since the homogeneous equation $(\mathcal{R}_C^+)$ has only the zero solution, $(\mathcal{R}_C^+)$ itself has a unique solution $\varphi \in C^{0,\alpha}(\partial S)$. The function u given by (1.35) belongs to $C^{1,\alpha}(\bar{S}^+)$ and $\Delta u = 0$ (in S^+), so u is the (unique) solution of problem (R$^+$).

(ii) Let φ_0 be a solution of the homogeneous equation $(\mathcal{R}_C^-)$, that is,

$$(W_0^* - \tfrac{1}{2}I)(\varphi_0 - (p\varphi_0)\Phi) - \sigma V_0(\varphi_0 - (p\varphi_0)\Phi) - \sigma(p\varphi_0) = 0.$$

Then the function

$$\mathcal{U}^- = \mathcal{V}^-(\varphi_0 - (p\varphi_0)\Phi) + p\varphi_0$$

satisfies

$$\Delta \mathcal{U}^- = 0,$$

$$\partial_\nu \mathcal{U}^- - \sigma \mathcal{U}^-|_{\partial S} = 0,$$

$$\mathcal{U}^-(x) = p(\varphi_0 - (p\varphi_0)\Phi)\ln|x| + p\varphi_0 + \mathcal{U}^{\mathcal{A}}(x) = \mathcal{U}^{\mathcal{A}}(x) + p\varphi_0 \quad \text{as } |x| \to \infty.$$

This is the homogeneous problem (R$^-$), so, again by Corollary 1.30(i),

$$p\varphi_0 = 0, \quad \mathcal{V}^-(\varphi_0 - (p\varphi_0)\Phi) + p\varphi_0 = \mathcal{V}^-\varphi_0 = 0,$$

from which

$$V_0^-\varphi_0 = V_0\varphi_0 = V_0^+\varphi_0 = 0.$$

Consequently, $\mathcal{V}^+\varphi_0 = 0$ as the unique solution of the homogeneous problem (D$^+$). From Remark 1.35(iii) we now conclude that $\varphi_0 = 0$. Hence, by the Fredholm

Alternative, $(\mathcal{R}_C^-)$ has a unique solution $\varphi \in C^{0,\alpha}(\partial S)$. As in (i), u defined by (1.36) satisfies $u \in C^{1,\alpha}(\bar{S}^-)$ and $\Delta u = 0$ (in S^-). Since $p(\varphi - (p\varphi)\Phi) = 0$, it follows that $u \in \mathcal{A}^*$, so u is the (unique) solution of (R^-).

1.8. The alternative indirect method

In the classical indirect method, the solutions of $(D^\pm)$ are sought in the form of double layer potentials, and those of $(N^\pm)$ in the form of single layer potentials. This gives rise to boundary equations where the integral operators have weakly singular kernels. In the "alternative" method, the choice of form for the solutions is the other way round. But when the solutions of $(N^\pm)$ are sought in the form of double layer potentials, the equations on ∂S for their densities have strongly singular operators, which is not helpful. Consequently, this method makes practical sense only for the Dirichlet problems, and we choose to solve (D^+) as an illustration.

We seek the solution of (D^+) as $u = \mathcal{V}^+\varphi$. This leads to the boundary integral equation

$$V_0\varphi = \mathcal{P}. \qquad (\mathcal{D}_A^+)$$

1.47. Theorem. *If $\omega \neq 0$, then $(\mathcal{D}_A^+)$ has a unique solution $\varphi \in C^{0,\alpha}(\partial S)$ for any $\mathcal{P} \in C^{1,\alpha}(\partial S)$, $\alpha \in (0,1)$.*

In this case,

$$u = \mathcal{V}^+\varphi$$

is the (unique) solution of (D^+).

Proof. Applying N_0 to both sides in $(\mathcal{D}_A^+)$ and using $(1.23)_2$, we find that any solution of $(\mathcal{D}_A^+)$ is also a solution of the equation

$$(W_0^{*2} - \tfrac{1}{4}I)\varphi = N_0\mathcal{P}. \qquad (1.38)$$

According to Theorem 1.39(i), the null space of the adjoint operator $W^2 - \tfrac{1}{4}I$ is spanned by 1. Since, by Lemma 1.44(iv),

$$(1, N_0\mathcal{P}) = p(N_0\mathcal{P}) = 0,$$

the Fredholm Alternative implies that (1.38) is solvable in $C^{0,\alpha}(\partial S)$ and that its solution is of the form $\varphi = \varphi_0 + a\Phi$, where φ_0 is any (fixed) solution and a is an arbitrary constant. Then we can write

$$0 = (W_0^{*2} - \tfrac{1}{4}I)(\varphi_0 + a\Phi) - N_0\mathcal{P} = N_0\big[V_0(\varphi_0 + a\Phi) - \mathcal{P}\big],$$

which, by Theorem 1.39(ii), means that

$$V_0(\varphi_0 + a\Phi) - \mathcal{P} = a' = \text{const}.$$

Since $V_0\Phi = \omega$, this yields

$$V_0\varphi_0 - \mathcal{P} = a' - a\omega = a'', \qquad (1.39)$$

where a'' is a constant that depends on the choice of φ_0. Now

$$V_0(\varphi_0 - a''\omega^{-1}\Phi) = V_0\varphi_0 - a'' = \mathcal{P};$$

therefore,

$$\varphi = \varphi_0 - a''\omega^{-1}\Phi = \varphi_0 - (V_0\varphi_0 - \mathcal{P})\omega^{-1}\Phi \qquad (1.40)$$

is a solution of $(\mathcal{D}_A^+)$ in $C^{0,\alpha}(\partial S)$. Its uniqueness follows from the fact that the difference $\varphi_1 - \varphi_2$ of any two solutions satisfies $V_0(\varphi_1 - \varphi_2) = 0$, which, according to Theorem 1.43, implies that $\varphi_1 - \varphi_2 = 0$.

Since $\mathcal{V}^+\varphi \in C^{1,\alpha}(\bar{S}^+)$ and $\Delta(\mathcal{V}^+\varphi) = 0$, it follows that $u = \mathcal{V}^+\varphi$ is the (unique) solution of (D^+), independent of the choice of φ_0 in its construction. This can also be verified directly. If instead of φ_0 we take another solution $\bar{\varphi}_0$ of (1.38), then from (1.39) we see that

$$V_0(\varphi_0 - \bar{\varphi}_0) = a'' - \bar{a}'',$$

and Corollary 1.41 yields

$$\varphi_0 = \bar{\varphi}_0 + k\Phi, \quad a'' = \bar{a}'' + k\omega$$

for some constant k. Hence, by (1.40),

$$\varphi = \varphi_0 - a''\omega^{-1}\Phi = \bar{\varphi}_0 + k\Phi - (\bar{a}'' + k\omega)\omega^{-1}\Phi = \bar{\varphi}_0 - \bar{a}''\omega^{-1}\Phi.$$

The situation needs different handling when $\omega = 0$. Thus, since here Φ cannot help us make the necessary adjustment of the density of $\mathcal{V}^+$ as above, we seek the solution in the form

$$u = \mathcal{V}^+\varphi + c, \quad c = \text{const}.$$

Then the corresponding boundary integral equation is

$$V_0\varphi = \mathcal{P} - c. \qquad (\tilde{\mathcal{D}}_A^+)$$

Applying q to both sides of this equality and using Lemma 1.44(i), we find that

$$0 = \omega(p\varphi) = q(V_0\varphi) = q\mathcal{P} - c(q1) = q\mathcal{P} - c(p\Phi) = q\mathcal{P} - c,$$

so

$$c = q\mathcal{P}. \tag{1.41}$$

1.48. Theorem. *If $\omega = 0$, then $(\tilde{\mathcal{D}}_A^+)$ with c given by (1.41) is solvable in $C^{0,\alpha}(\partial S)$ for any $\mathcal{P} \in C^{1,\alpha}(\partial S)$, and its solution is unique up to a term of the form $a\Phi$, where a is an arbitrary constant.*

In this case, the (unique) solution of (D^+) *is*

$$u = \mathcal{V}^+(\varphi) + q\mathcal{P},$$

where φ is any solution of $(\tilde{\mathcal{D}}_A^+)$.

Proof. As before, we apply N_0 to $(\tilde{\mathcal{D}}_A^+)$. Since, by Theorem 1.39(ii), $N_0 c = 0$, we again arrive at equation (1.38). But now $V_0\Phi = \omega = 0$, so (1.39) is replaced by

$$V_0\varphi_0 - \mathcal{P} = a' = \mathrm{const},$$

where φ_0 is any solution of (1.38). Applying q once more, we find that $a' = -q\mathcal{P}$, which means that φ_0 is also a solution of $(\tilde{\mathcal{D}}_A^+)$; in other words, the sets of solutions of (1.38) and $(\tilde{\mathcal{D}}_A^+)$ coincide. Consequently, the solution of $(\tilde{\mathcal{D}}_A^+)$ is unique up to a term $a\Phi$, $a = \mathrm{const}$.

Since for any solution $\varphi \in C^{0,\alpha}(\partial S)$ of $(\tilde{\mathcal{D}}_A^+)$ we have $\mathcal{V}^+\varphi \in C^{1,\alpha}(\bar{S}^+)$ and $\Delta(\mathcal{V}^+\varphi + q\mathcal{P}) = 0$, the function $u = \mathcal{V}^+\varphi + q\mathcal{P}$ is the (unique) solution of (D^+). The arbitrariness $a\Phi$ in φ is eliminated by the fact that in this case $\mathcal{V}^+\Phi = \omega = 0$.

1.9. The modified indirect method

The drawback of a non-unique solution for $(\mathcal{D}_A^+)$ can be eliminated. Consider a different fundamental solution for $-\Delta$, namely,

$$g^c(x, y) = g(x, y) + c, \quad c = \mathrm{const}, \tag{1.42}$$

and the corresponding modified single layer potential

$$(\mathcal{V}^c\varphi)(x) = \int_{\partial S} \big(g(x, y) + c\big)\varphi(y)\, ds(y) = (\mathcal{V}\varphi)(x) + c(p\varphi). \tag{1.43}$$

Since the double layer potential remains unchanged under (1.42), we do not need to append the superscript c to its symbol. The same is obviously valid for the boundary operators W_0^* and N_0.

In what follows we use the notation $\mathcal{V}^{c\pm}$, $V_0^{c\pm}$ and V_0^c with the obvious meaning.

1.49. Theorem. *If $\varphi \in C^{0,\alpha}(\partial S)$, $\alpha \in (0,1)$, $c \neq -\omega$ and $V_0^c \varphi = 0$, then $\varphi = 0$.*

Proof. For $\varphi \in C^{0,\alpha}(\partial S)$, we define the function

$$\mathcal{U}^- = \mathcal{V}^{c-}\varphi - (p\varphi)(\mathcal{V}^-\Phi) + \omega(p\varphi).$$

From (1.43), (1.14), (1.32) and the assumption that $V_0^c \varphi = 0$ we see that $\mathcal{U}^-$ is a solution of the homogeneous exterior Dirichlet problem

$$\Delta \mathcal{U}^- = 0,$$
$$\mathcal{U}^-|_{\partial S} = V_0^c \varphi - (p\varphi)(V_0\Phi) + \omega(p\varphi) = 0, \qquad (1.44)$$
$$\mathcal{U}^-(x) = -(2\pi)^{-1}(p\varphi)\ln|x| + c(p\varphi) + (2\pi)^{-1}(p\Phi)(p\varphi)\ln|x| + \omega(p\varphi) + \mathcal{U}^{\mathcal{A}}(x)$$
$$= \mathcal{U}^{\mathcal{A}}(x) + (c+\omega)(p\varphi) \quad \text{as } |x| \to \infty.$$

By Corollary 1.30(i), we then must have $(c+\omega)(p\varphi) = 0$, and, since $c \neq -\omega$, we deduce that

$$p\varphi = 0. \qquad (1.45)$$

Thus, $\mathcal{V}^{c-}\varphi = \mathcal{V}^-\varphi$, and from (1.44) and (1.45) it follows that $\mathcal{V}^-\varphi \in \mathcal{A}$ is the solution of the homogeneous problem (D$^-$); therefore, $\mathcal{V}^-\varphi = 0$. Then $V_0\varphi = 0$, so $\mathcal{V}^+\varphi$ is the solution of the homogeneous problem (D$^+$), which means that $\mathcal{V}^+\varphi = 0$. We now use Remark 1.35(iii) to deduce that $\varphi = 0$.

1.50. Theorem. *The composition relations (1.23) and (1.24) remain valid if V_0 is replaced by V_0^c.*

Proof. By (1.42),

$$V_0\varphi = V_0^c\varphi - c(p\varphi),$$

which, replaced in (1.23) and (1.24)$_2$ written for $\varphi \in C^{0,\alpha}(\partial S)$ and $\psi \in C^{1,\alpha}(\partial S)$, respectively, yields

$$W_0\bigl(V_0^c\varphi - c(p\varphi)\bigr) = V_0^c(W_0^*\varphi) - cp(W_0^*\varphi),$$
$$N_0\bigl(V_0^c\varphi - c(p\varphi)\bigr) = (W_0^{*2} - \tfrac{1}{4}I)\varphi, \qquad (1.46)$$
$$V_0^c(N_0\psi) - cp(N_0\psi) = (W_0^2 - \tfrac{1}{4}I)\psi.$$

Since, by Remark 1.35(v),
$$W_0(c(p\varphi)) = c(p\varphi)(W_0 1) = -\tfrac{1}{2}c(p\varphi),$$
by Lemma 1.44(ii)
$$cp(W_0^*\varphi) = -\tfrac{1}{2}c(p\varphi),$$
by Theorem 1.39(ii)
$$N_0(c(p\varphi)) = c(p\varphi)(N_0 1) = 0,$$
and by Lemma 1.44(iv)
$$cp(N_0\psi) = 0,$$
the desired equalities follow immediately from (1.46).

For simplicity, in this section we continue to quote the composition formulae as (1.23) and (1.24), but we understand them with V_0^c instead of V_0.

We now seek the solution of (D^+) in the form $u = \mathcal{V}^{c+}\varphi$ with $c \neq -\omega$ chosen a priori, so the problem reduces to the solution of the Fredholm equation of the first kind
$$V_0^c \varphi = \mathcal{P}. \tag{$\mathcal{D}_M^+$}$$

1.51. Theorem. *If $\mathcal{P} \in C^{1,\alpha}(\partial S)$, $\alpha \in (0,1)$, then $(\mathcal{D}_M^+)$ has a unique solution $\varphi \in C^{0,\alpha}(\partial S)$. In this case,*
$$u = \mathcal{V}^{c+}\varphi$$
is the (unique) solution of (D^+).

Proof. We follow the general scheme used in the proof of Theorem 1.47.

Applying N_0 to $(\mathcal{D}_M^+)$, in view of $(1.23)_2$ we find that any solution of $(\mathcal{D}_M^+)$ is also a solution of the equation
$$(W_0^{*2} - \tfrac{1}{4}I)\varphi = N_0 \mathcal{P}. \tag{1.47}$$

Since the null space of the operator of the adjoint equation is spanned by 1 and, by Lemma 1.44(iv),
$$(1, N_0 \mathcal{P}) = p(N_0 \mathcal{P}) = 0,$$
the Fredholm Alternative tells us that (1.47) is solvable in $C^{0,\alpha}(\partial S)$ and its solution is $\varphi = \varphi_0 + a\Phi$, where φ_0 is any (fixed) solution of (1.47) and a is an arbitrary constant. Then we can write
$$0 = (W_0^{*2} - \tfrac{1}{4}I)(\varphi_0 + a\Phi) - N_0 \mathcal{P} = N_0(V_0^c(\varphi_0 + a\Phi) - \mathcal{P}),$$

which, by Theorem 1.39(ii), implies that

$$V_0^c(\varphi_0 + a\Phi) - \mathcal{P} = a' = \text{const}.$$

Taking (1.32) into account, we rewrite this in the form

$$a' = V_0^c\varphi_0 + a(V_0\Phi + c(p\Phi)) - \mathcal{P} = V_0^c\varphi_0 + a(c + \omega) - \mathcal{P};$$

hence,

$$V_0^c\varphi_0 - \mathcal{P} = V_0\varphi_0 + c(p\varphi_0) - \mathcal{P} = a' - a(\omega + c) = a''.$$

Using q and Lemma 1.44(i), we find that

$$\omega(p\varphi_0) + c(p\varphi_0) - q\mathcal{P} = (\omega + c)(p\varphi_0) - q\mathcal{P} = a''.$$

Now

$$V_0^c\big(\varphi_0 - a''(c+\omega)^{-1}\Phi\big) = \mathcal{P} + a'' - a''(c+\omega)^{-1}(V_0\Phi + c(p\Phi))$$
$$= \mathcal{P} + a'' - a''(c+\omega)^{-1}(\omega + c) = \mathcal{P},$$

so

$$\varphi = \varphi_0 - a''(c+\omega)^{-1}\Phi = \varphi_0 + \big((c+\omega)^{-1}(q\mathcal{P}) - p\varphi_0\big)\Phi \in C^{0,\alpha}(\partial S)$$

is a solution of $(\mathcal{D}_M^+)$.

To show that this solution is unique, we note that the difference $\varphi_1 - \varphi_2$ of two solutions satisfies $V_0^c(\varphi_1 - \varphi_2) = 0$; by Theorem 1.49, this yields $\varphi_1 = \varphi_2$.

The function

$$u = \mathcal{V}^{c+}\varphi = \mathcal{V}^{c+}\big(\varphi_0 + \big((c+\omega)^{-1}(q\mathcal{P}) - p\varphi_0\big)\Phi\big)$$

satisfies $u \in C^{1,\alpha}(\bar{S}^+)$ and $\Delta u = 0$ (in S^+), so it is the (unique) solution of (D^+).

1.52. Remark. Clearly, the representation of the solution of (D^+) as $u = \mathcal{V}^{c+}\varphi$ is not unique, since the density φ depends on the choice of the constant c. However, u itself is unique, so for any two distinct constants c_1 and c_2, $c_1 \neq -\omega$, $c_2 \neq -\omega$, the corresponding densities φ_1 and φ_2 must satisfy

$$V_0^{c_1}\varphi_1 = V_0^{c_2}\varphi_2 = \mathcal{P}.$$

By (1.43), this means that

$$V_0(\varphi_1 - \varphi_2) = c_2(p\varphi_2) - c_1(p\varphi_1) = \text{const};$$

hence, according to Corollary 1.41,

$$\varphi_1 - \varphi_2 = k\Phi, \quad c_2(p\varphi_2) - c_1(p\varphi_1) = k\omega, \quad k = \text{const}, \tag{1.48}$$

so

$$p\varphi_1 - p\varphi_2 = k(p\Phi) = k. \tag{1.49}$$

From $(1.48)_2$ and (1.49) it follows that

$$-\omega(p\varphi_1 - p\varphi_2) = c_1(p\varphi_1) - c_2(p\varphi_2),$$

which yields

$$p\varphi_2 = (c_1 + \omega)(c_2 + \omega)^{-1}(p\varphi_1).$$

Using (1.49) again, we now find that

$$k = \left[1 - (c_1 + \omega)(c_2 + \omega)^{-1}\right](p\varphi_1) = (c_2 - c_1)(c_2 + \omega)^{-1}(p\varphi_1);$$

therefore, by $(1.48)_1$,

$$\varphi_2 = \varphi_1 + (c_1 - c_2)(c_2 + \omega)^{-1}(p\varphi_1)\Phi. \tag{1.50}$$

Equality (1.50) shows how any solution $\varphi = \varphi(c)$ of the integral equation $(\mathcal{D}_M^+)$ can be generated once the solution for a particular value of $c \neq -\omega$ has been found.

The modified indirect method enables us to solve an exterior problem that is more general than (D^-). Thus, we consider the boundary value problem

$$\begin{aligned}
(\Delta u)(x) &= 0, \quad x \in S^-, \\
u(x) &= \mathcal{R}(x), \quad x \in \partial S, \\
u(x) &= s \ln |x| + u^{\mathcal{A}^*}(x) \quad \text{as } |x| \to \infty,
\end{aligned} \tag{D_G^-}$$

where s is a prescribed constant. Clearly, (D^-) corresponds to $s = 0$.

1.53. Theorem. *If $\mathcal{R} \in C^{1,\alpha}(\partial S)$, $\alpha \in (0,1)$, then (D_G^-) has a unique solution, which can be expressed as a modified single layer potential.*

Proof. By Theorem 1.51, there is a density $\varphi \in C^{0,\alpha}(\partial S)$ such that $\mathcal{V}^{c+}\varphi$ (with $c \neq -\omega$ chosen a priori) satisfies

$$\mathcal{V}_0^{c+}\varphi = \mathcal{V}_0^c\varphi = \mathcal{R}. \tag{1.51}$$

Since $\mathcal{V}_0^{c-}\varphi = \mathcal{V}_0^c\varphi$, it follows that the function $u = \mathcal{V}^{c-}\varphi$, $c \neq -\omega$, satisfies the first two relations in (D_G^-). However, the third one might not hold for this function, so we modify the procedure and take

$$u = \mathcal{V}^{c'-}\varphi', \quad c' = \text{const}, \quad c' \neq -\omega, \quad \varphi' = \varphi + a\Phi, \quad a = \text{const},$$

where φ is the density mentioned above. It is obvious that $u \in C^{1,\alpha}(\bar{S}^-)$.

Suppose first that $s \neq 0$. By (1.32),

$$\Delta(\mathcal{V}^{c'-}\varphi') = 0,$$
$$\mathcal{V}_0^{c'-}\varphi' = V_0\varphi + c'(p\varphi) + a\big(V_0\Phi + c'(p\Phi)\big)$$
$$= V_0^c\varphi + (c'-c)(p\varphi) + a(\omega + c') = \mathcal{R},$$

if $(c'-c)(p\varphi) + a(\omega + c') = 0$, that is, if

$$c' = \big(c(p\varphi) - a\omega\big)(p\varphi + a)^{-1}, \tag{1.52}$$

and

$$(\mathcal{V}^{c'-}\varphi')(x) = (\mathcal{V}^-\varphi)(x) + a(\mathcal{V}^-\Phi)(x) + c'(p\varphi')$$
$$= -(2\pi)^{-1}(p\varphi + a)\ln|x| + \mathcal{U}^{\mathcal{A}^*}(x)$$
$$= s\ln|x| + \mathcal{U}^{\mathcal{A}^*}(x) \quad \text{as } |x| \to \infty$$

if

$$a = -(p\varphi + 2\pi s). \tag{1.53}$$

From (1.52) and (1.53) we obtain

$$c' = -\omega - (2\pi s)^{-1}(c+\omega)(p\varphi). \tag{1.54}$$

Consequently, if $s \neq 0$, then the (unique) solution of (D_G^-) is

$$u = \mathcal{V}^{c'-}\big(\varphi - (p\varphi + 2\pi s)\Phi\big),$$

with c' given by (1.54).

When $s = 0$, the solution of (D_G^-) (which now becomes (D^-)), is unobtainable as above. In this case (1.53) yields $a = -p\varphi$, and we have

$$\mathcal{V}_0^{c'-}\varphi' = V_0^c\varphi + (c'-c)(p\varphi) - (\omega + c')(p\varphi) = \mathcal{R} - (c+\omega)(p\varphi)$$

for any $c' = \text{const}$. Hence, to satisfy the boundary condition with $\mathcal{V}^{c'-}\varphi'$ we need $(c+\omega)(p\varphi) = 0$. But $c+\omega \neq 0$ and we cannot guarantee that $p\varphi = 0$. We recall that φ is the $C^{0,\alpha}$-density satisfying $V_0^c \varphi = \mathcal{R}$, or $V_0\varphi + c(p\varphi) = \mathcal{R}$. Applying q to both sides and using Lemma 1.44(i), we find that

$$q(V_0\varphi) + c(p\varphi)(q1) = \omega(p\varphi) + c(p\varphi)(p\Phi) = (\omega + c)(p\varphi) = q\mathcal{R},$$

and $\mathcal{R}$ may not be such that $q\mathcal{R} = 0$. However, since, by (1.51) and (1.43),

$$V_0^c(\varphi - (p\varphi)\Phi) + (c+\omega)(p\varphi) = \mathcal{R} - (p\varphi)(V_0\Phi + c(p\Phi)) + (c+\omega)(p\varphi) = \mathcal{R}$$

and $p(\varphi - (p\varphi)\Phi) = 0$ implies that

$$\mathcal{V}^{c-}(\varphi - (p\varphi)\Phi) + (c+\omega)(p\varphi) \in \mathcal{A}^*,$$

it follows that now the (unique) solution of (D$^-$) is

$$u = \mathcal{V}^{c-}(\varphi - (p\varphi)\Phi) + (c+\omega)(p\varphi).$$

1.10. The refined indirect method

In the modified indirect method we need to know ω so that we can choose $c \neq -\omega$. To avoid this, we seek the solution of (D$^+$) in the form

$$u = \mathcal{V}^+\varphi - c, \tag{1.55}$$

where φ satisfies $p\varphi = s$ and c and s are constants, with s chosen a priori. Using the boundary condition, we then obtain the system of boundary integral equations

$$V_0\varphi - c = \mathcal{P}, \quad p\varphi = s. \tag{$\mathcal{D}_R^+$}$$

1.54. Theorem. ($\mathcal{D}_R^+$) *has a unique solution* (φ, c) *with* $\varphi \in C^{0,\alpha}(\partial S)$ *for any* $\mathcal{P} \in C^{1,\alpha}(\partial S)$, $\alpha \in (0,1)$.

Then the (unique) solution of (D$^+$) *is given by* (1.55).

Proof. Applying the operator N_0 to both sides of the first equation in ($\mathcal{D}_R^+$) and using the composition formula (1.23)$_2$ and Theorem 1.39(ii), we see that any solution of ($\mathcal{D}_R^+$) is also a solution of the Fredholm equation of the second kind

$$(W_0^{*2} - \tfrac{1}{4}I)\varphi = N_0\mathcal{P}. \tag{1.56}$$

By Lemma 1.44(iv), $p(N_0\mathcal{P}) = 0$, so (1.56) is solvable in $C^{0,\alpha}(\partial S)$. Since the null space of $W_0^{*2} - \frac{1}{4}I$ is spanned by Φ (Theorem 1.39(i)), we write the solution as

$$\varphi = \varphi_0 + k\Phi, \tag{1.57}$$

where φ_0 is any (fixed) solution of (1.56) and k is an arbitrary constant. Then

$$(W_0^{*2} - \tfrac{1}{4}I)(\varphi_0 + k\Phi) = N_0\mathcal{P},$$

from which, by applying the operator V_0 to both sides and using $(1.23)_1$ and $(1.24)_2$, we obtain

$$(W_0^2 - \tfrac{1}{4}I)\big(V_0(\varphi_0 + k\Phi) - \mathcal{P}\big) = 0.$$

Since the null space of $W_0^2 - \frac{1}{4}I$ is spanned by 1, this implies that

$$V_0(\varphi_0 + k\Phi) - \mathcal{P} = c, \tag{1.58}$$

where, for φ_0 fixed, the constant c depends only on k.

Recalling that $p\Phi = 1$, from (1.57) we have

$$p\varphi = p(\varphi_0 + k\Phi) = p\varphi_0 + k,$$

so $p\varphi = s$ if

$$k = s - p\varphi_0. \tag{1.59}$$

At the same time, applying q to both sides in (1.58) and using Lemma 1.44(i), we obtain

$$wp(\varphi_0 + k\Phi) - q\mathcal{P} = c(q1) = c(p\Phi) = c. \tag{1.60}$$

Replacing k from (1.59) in (1.57) and (1.60), we now conclude that a solution (φ, c) of $(\mathcal{D}_R^+)$ is given by

$$\varphi = \varphi_0 + (s - p\varphi_0)\Phi, \quad c = ws - q\mathcal{P}. \tag{1.61}$$

To show that the solution of $(\mathcal{D}_R^+)$ is unique, we note that the difference $(\bar{\varphi}, \bar{c})$ of two solutions satisfies

$$V_0\bar{\varphi} - \bar{c} = 0, \quad p\bar{\varphi} = 0.$$

Consequently, $\mathcal{V}^-\bar{\varphi} \in \mathcal{A}$, and $\mathcal{V}^-\bar{\varphi} - \bar{c}$ is the solution of the homogeneous problem (D$^-$). By Corollary 1.30(i),
$$\bar{c} = 0, \quad \mathcal{V}^-\bar{\varphi} = 0.$$

Thus, $V_0\bar{\varphi} = 0$, so $\mathcal{V}^+\bar{\varphi}$ is the unique solution of the homogeneous problem (D$^+$). Consequently, $\mathcal{V}^+\bar{\varphi} = 0$, and from Remark 1.35(iii) it now follows that $\bar{\varphi} = 0$.

Since $\mathcal{V}^+\varphi \in C^{1,\alpha}(\bar{S}^+)$ and $\Delta(\mathcal{V}^+\varphi - c) = 0$, it follows that $\mathcal{V}^+\varphi - c$ is the (unique) solution of (D$^+$).

1.55. Remark. Again, the representation $u = \mathcal{V}^+(\varphi) - c$ is not unique. If (φ_1, c_1) and (φ_2, c_2) are the solutions of $(\mathcal{D}_R^+)$ for $s_1 \neq s_2$, then from (1.61) we see that

$$\varphi_2 = \varphi_1 + (s_2 - s_1)\Phi, \quad c_2 = c_1 + \omega(s_2 - s_1). \tag{1.62}$$

We now turn our attention to the exterior problem (D$_G^-$) discussed in §1.9.

1.56. Theorem. *(D$_G^-$) has a unique solution for any $\mathcal{R} \in C^{1,\alpha}(\partial S)$, $\alpha \in (0,1)$, which is constructed with the same density and additive constant as in (D$^+$) with $\mathcal{P}$ replaced by $\mathcal{R}$ and with s taken from the far-field pattern of the problem.*

Proof. The function $\mathcal{V}^-\varphi - c$, where (φ, c) is the unique solution of $(\mathcal{D}_R^+)$ with $\mathcal{P}$ replaced by $\mathcal{R}$, satisfies
$$\Delta(\mathcal{V}^-\varphi - c) = 0,$$
$$V_0^-\varphi - c = V_0\varphi - c = \mathcal{R},$$
$$(\mathcal{V}^-\varphi - c)(x) = (p\varphi)\ln|x| + \mathcal{U}^{\mathcal{A}}(x) - c = s\ln|x| + \mathcal{U}^{\mathcal{A}^*}(x) \quad \text{as } |x| \to \infty.$$

Hence, $\mathcal{V}^-\varphi - c \in C^{1,\alpha}(\bar{S}^-)$ is a solution of (D$_G^-$). This solution is unique since the difference of any two solutions satisfies the homogeneous problem (D$^-$).

1.57. Remarks. (i) Theorem 1.56 illustrates an unusual but very useful feature of the indirect refined method: the solutions of both (D$^+$) and (D$_G^-$) are given by one and the same formula.

(ii) This method also enables us to determine ω and Φ. Let (φ_1, c_1) and (φ_2, c_2) be the solutions of $(\mathcal{D}_R^+)$ corresponding to two numbers $s_1 \neq s_2$ and any suitably smooth data $\mathcal{P}$. Then, by (1.62),
$$\varphi_1 - \varphi_2 = (s_1 - s_2)\Phi, \quad c_1 - c_2 = (s_1 - s_2)\omega;$$
hence,
$$\Phi = (s_1 - s_2)^{-1}(\varphi_1 - \varphi_2), \quad \omega = (s_1 - s_2)^{-1}(c_1 - c_2).$$

1.11. The direct method

The direct method is based on Green's representation formulae. As in the case of the classical indirect method, first we discuss the Dirichlet and Neumann boundary value problems.

For a harmonic function u in S^+, the representation formula $(1.19)_1$ restricted to ∂S can be written as

$$V_0(\partial_\nu u) - (W_0 + \tfrac{1}{2}I)(u|_{\partial S}) = 0.$$

Then the boundary integral equations for (D^+) and (N^+) are

$$V_0 \varphi = (W_0 + \tfrac{1}{2}I)\mathcal{P}, \tag{$\mathcal{D}_D^+$}$$

$$(W_0 + \tfrac{1}{2}I)\psi = V_0 \mathcal{Q}, \tag{$\mathcal{N}_D^+$}$$

where $\varphi = \partial_\nu u$ and $\psi = u|_{\partial S}$, respectively.

The representation formula $(1.20)_1$ for a harmonic function u in S^- requires that $u \in \mathcal{A}$, so it can be used only for (N^-). Thus, writing it restricted to ∂S in the form

$$-V_0(\partial_\nu u) + (W_0 - \tfrac{1}{2}I)(u|_{\partial S}) = 0,$$

for (N^-) we arrive at the boundary integral equation

$$(W_0 - \tfrac{1}{2}I)\psi = V_0 \mathcal{S}. \tag{$\mathcal{N}_D^-$}$$

For the solution of (D^-) we need a different representation formula, which is valid for a harmonic function $u \in \mathcal{A}^*$ in S^-. Since in this case $u = u^\mathcal{A} + c$, where $c = \text{const}$, we apply $(1.20)_1$ to $u^\mathcal{A} = u - c \in \mathcal{A}$ and find that

$$u - c = -\mathcal{V}^-(\partial_\nu(u - c)) + \mathcal{W}^-((u - c)|_{\partial S}).$$

By Remark 1.35(v), we have $\mathcal{W}^- c = 0$, so

$$u = -\mathcal{V}^-(\partial_\nu u) + \mathcal{W}^-(u|_{\partial S}) + c. \tag{1.63}$$

This equality restricted to ∂S yields

$$-V_0(\partial_\nu u) + (W_0 - \tfrac{1}{2}I)(u|_{\partial S}) + c = 0;$$

THE DIRECT METHOD

therefore, the boundary integral equation for (D$^-$) is

$$V_0\varphi = (W_0 - \tfrac{1}{2}I)\mathcal{R} + c. \qquad (\mathcal{D}_D^-)$$

The value of c is determined in the process of solution.

In this section we consider only $(\mathcal{D}_D^+)$ and $(\mathcal{D}_D^-)$. The integral equations $(\mathcal{N}_D^+)$ and $(\mathcal{N}_D^-)$ are solved in §1.12.

Since the unknown function φ in $(\mathcal{D}_D^+)$ and $(\mathcal{D}_D^-)$ represents the normal derivative on ∂S of the solution u of the corresponding Dirichlet problem, Theorem 1.24 shows that φ must also satisfy

$$p\varphi = 0. \qquad (1.64)$$

In the case of $(\mathcal{D}_D^-)$, this condition also ensures that the solution of (D$^-$) belongs to $\mathcal{A}^*$, as required.

1.58. Theorem. (i) *For any $\mathcal{P} \in C^{1,\alpha}(\partial S)$, $\alpha \in (0,1)$, the pair of equations $(\mathcal{D}_D^+)$ and (1.64) has a unique solution $\varphi \in C^{0,\alpha}(\partial S)$.*

Then (D$^+$) *has the (unique) solution*

$$u = \mathcal{V}^+\varphi - \mathcal{W}^+\mathcal{P}. \qquad (1.65)$$

(ii) *For any $\mathcal{R} \in C^{1,\alpha}(\partial S)$, $\alpha \in (0,1)$, the pair of equations $(\mathcal{D}_D^-)$ with $c = q\mathcal{R}$ and (1.64) has a unique solution $\varphi \in C^{0,\alpha}(\partial S)$.*

Then (D$^-$) *has the (unique) solution*

$$u = -\mathcal{V}^-\varphi + \mathcal{W}^-\mathcal{R} + q\mathcal{R}. \qquad (1.66)$$

Proof. (i) Applying N_0 to both sides in $(\mathcal{D}_D^+)$ and using $(1.23)_2$ and $(1.24)_1$, we see that any solution of $(\mathcal{D}_D^+)$ is also a solution of

$$(W_0^{*2} - \tfrac{1}{4}I)\varphi = (W_0^* + \tfrac{1}{2}I)(N_0\mathcal{P}). \qquad (1.67)$$

This equation is solvable in $C^{0,\alpha}(\partial S)$ since the null space of the operator of the adjoint equation is spanned by 1 (Theorem 1.39(i)) and, by Lemma 1.44(ii),

$$\bigl(1, (W_0^* + \tfrac{1}{2}I)(N_0\mathcal{P})\bigr) = p(W_0^* + \tfrac{1}{2}I)(N_0\mathcal{P}) = -\tfrac{1}{2}p(N_0\mathcal{P}) + \tfrac{1}{2}p(N_0\mathcal{P}) = 0.$$

The general solution of (1.67) is

$$\varphi = \varphi_0 + a\Phi, \qquad (1.68)$$

where φ_0 is a (fixed) solution and $a = $ const is arbitrary. Thus,

$$(W_0^{*2} - \tfrac{1}{4}I)(\varphi_0 + a\Phi) = (W_0^* + \tfrac{1}{2}I)(N_0\mathcal{P}),$$

which, after operation with V_0, becomes

$$(W_0^2 - \tfrac{1}{4}I)\bigl[V_0(\varphi_0 + a\Phi) - (W_0 + \tfrac{1}{2}I)\mathcal{P}\bigr] = 0.$$

Consequently, by Theorem 1.39(i),

$$V_0(\varphi_0 + a\Phi) - (W_0 + \tfrac{1}{2}I)\mathcal{P} = a' = \text{const}. \tag{1.69}$$

Next, from (1.68) we find that

$$p\varphi = p\varphi_0 + a(p\Phi) = p\varphi_0 + a,$$

so the only function (1.68) that also satisfies (1.64) is

$$\varphi = \varphi_0 - (p\varphi_0)\Phi. \tag{1.70}$$

Since $V_0\Phi = \omega$, equality (1.69) with $a = -p\varphi_0$ can be rewritten as

$$V_0\varphi_0 - (W_0 + \tfrac{1}{2}I)\mathcal{P} = a' + \omega(p\varphi_0). \tag{1.71}$$

Applying q above and taking Lemma 1.44(i),(iii) into account, we see that

$$\omega(p\varphi_0) + \tfrac{1}{2}q\mathcal{P} - \tfrac{1}{2}q\mathcal{P} = a' + \omega(p\varphi_0),$$

or $a' = 0$. Then (1.69) reduces to $(\mathcal{D}_D^+)$, which means that, for any solution φ_0 of equation (1.67), the function (1.70) is a solution of both $(\mathcal{D}_D^+)$ and (1.64).

The difference $\bar{\varphi}$ of two such solutions satisfies

$$V_0\bar{\varphi} = 0, \quad p\bar{\varphi} = 0.$$

By Corollary 1.41, there is $k = $ const such that $\bar{\varphi} = k\Phi$, and $p\bar{\varphi} = k(p\Phi) = k = 0$ implies that $\bar{\varphi} = 0$; in other words, the pair $(\mathcal{D}_D^+)$ and (1.64) has a unique solution.

To verify that u given by (1.65) with φ as in (1.70) is the (unique) solution of (D^+), we note that, since $\varphi \in C^{0,\alpha}(\partial S)$ and $\mathcal{P} \in C^{1,\alpha}(\partial S)$, it follows that $u \in C^{1,\alpha}(\bar{S}^+)$. We also have $Au = 0$ (in S^+). Finally, restricting (1.65) to ∂S and recalling that φ is a solution of $(\mathcal{D}_D^+)$, we obtain

$$u|_{\partial S} = V_0\varphi - (W_0 - \tfrac{1}{2}I)\mathcal{P} = \bigl[V_0\varphi - (W_0 + \tfrac{1}{2}I)\mathcal{P}\bigr] + \mathcal{P} = \mathcal{P};$$

THE DIRECT METHOD 43

that is, u satisfies the required boundary condition.

(ii) Using arguments similar to those in (i), we find that any solution of $(\mathcal{D}_D^-)$ is also a solution of

$$(W_0^{*2} - \tfrac{1}{4}I)\varphi = (W_0^* - \tfrac{1}{2}I)(N_0\mathcal{R}). \tag{1.72}$$

Since, by Lemma 1.44(ii),(iv),

$$\big(1, (W_0^* - \tfrac{1}{2}I)(N_0\mathcal{R})\big) = p\big((W_0^* - \tfrac{1}{2}I)(N_0\mathcal{R})\big) = -\tfrac{1}{2}p(N_0\mathcal{R}) - \tfrac{1}{2}p(N_0\mathcal{R}) = 0,$$

equation (1.72) is solvable in $C^{0,\alpha}(\partial S)$ and its general solution is again of the form (1.68). From (1.64) we deduce once more that the function (1.68) satisfying both $(\mathcal{D}_D^-)$ and (1.64) is given by $a = -p\varphi_0$.

In place of (1.69) and (1.71) this time we obtain

$$V_0\big(\varphi_0 - (p\varphi_0)\Phi\big) - (W_0 - \tfrac{1}{2}I)\mathcal{R} = a', \tag{1.73}$$

$$V_0\varphi_0 - (W_0 - \tfrac{1}{2}I)\mathcal{R} = a' + \omega(p\varphi_0). \tag{1.74}$$

Applying q in (1.74) leads to

$$\omega(p\varphi_0) + \tfrac{1}{2}q\mathcal{R} + \tfrac{1}{2}q\mathcal{R} = a' + \omega(p\varphi_0),$$

from which $a' = q\mathcal{R}$. Consequently, (1.73) reduces to $(\mathcal{D}_D^-)$ with $c = q\mathcal{R}$. The uniqueness of the solution (1.70) of $(\mathcal{D}_D^-)$ and (1.64) is shown as in (i).

Since u defined by (1.66) belongs to $C^{1,\alpha}(\bar{S}^-)$, $\Delta u = 0$ (in S^-) and, as follows from $(\mathcal{D}_D^-)$,

$$u|_{\partial S} = -V_0\varphi + (W_0 + \tfrac{1}{2}I)\mathcal{R} + q\mathcal{R} = \big[-V_0\varphi + (W_0 - \tfrac{1}{2}I)\mathcal{R} + q\mathcal{R}\big] + \mathcal{R} = \mathcal{R},$$

we conclude that u is the (unique) solution of (D^-).

We turn our attention to the Robin problems. The boundary condition for (R^+) can be written as

$$\partial_\nu u = -\sigma u|_{\partial S} + \mathcal{K}, \tag{1.75}$$

which, replaced in the representation formula $(1.19)_1$, yields

$$u = -\mathcal{V}^+(\sigma u|_{\partial S}) - \mathcal{W}^+(u|_{\partial S}) + \mathcal{V}^+\mathcal{K}; \tag{1.76}$$

restricted to ∂S, this equality becomes

$$u|_{\partial S} = -V_0(\sigma u|_{\partial S}) - (W_0 - \tfrac{1}{2}I)(u|_{\partial S}) + V_0\mathcal{K}. \tag{1.77}$$

Setting $\varphi = u|_{\partial S}$, we rewrite (1.77) in the form

$$V_0(\sigma\varphi) + (W_0 + \tfrac{1}{2}I)\varphi = V_0\mathcal{K}. \qquad (\mathcal{R}_{\mathrm{D}}^+)$$

Since the solution of (R^-) belongs to $\mathcal{A}^*$, we seek it in the form

$$u = u^{\mathcal{A}} + c, \quad c = \mathrm{const}.$$

Then the representation formula (1.63) with

$$\partial_\nu u = \sigma u|_{\partial S} + \mathcal{L} \qquad (1.78)$$

can be written as

$$u = -\mathcal{V}^-(\sigma u|_{\partial S} + \mathcal{L}) + \mathcal{W}^-(u|_{\partial S}) + c. \qquad (1.79)$$

To have $u \in \mathcal{A}^*$, we need $p(\sigma u|_{\partial S} + \mathcal{L}) = 0$. With $u|_{\partial S} = \varphi$, (1.79) becomes

$$V_0(\sigma\varphi) - (W_0 - \tfrac{1}{2}I)\varphi - c = -V_0\mathcal{L}. \qquad (\mathcal{R}_{\mathrm{D}}^-)$$

Just as for $(\mathrm{D}^\pm)$, here the unknown densities φ must satisfy a second condition. Writing the boundary conditions (1.75) and (1.78) as

$$\partial_\nu u = -\sigma\varphi + \mathcal{K},$$
$$\partial_\nu u = \sigma\varphi + \mathcal{L},$$

applying p to both sides and taking into account that $p(\partial_\nu u) = 0$, we obtain, respectively,

$$p(\sigma\varphi - \mathcal{K}) = 0, \qquad (1.80)$$
$$p(\sigma\varphi + \mathcal{L}) = 0. \qquad (1.81)$$

1.59. Theorem. *Suppose that $\sigma \in C^{0,\alpha}(\partial S)$, $\alpha \in (0,1)$, and $\omega \neq 0$.*

(i) *The pair of equations $(\mathcal{R}_{\mathrm{D}}^+)$ and (1.80) has a unique solution $\varphi \in C^{1,\alpha}(\partial S)$ for any $\mathcal{K} \in C^{0,\alpha}(\partial S)$.*

Then (R^+) has the (unique) solution

$$u = -\mathcal{V}^+(\sigma\varphi) - \mathcal{W}^+\varphi + \mathcal{V}^+\mathcal{K}.$$

(ii) *The pair of equations $(\mathcal{R}_{\mathrm{D}}^-)$ with $c = q\varphi$ and (1.81) has a unique solution $\varphi \in C^{1,\alpha}(\partial S)$ for any $\mathcal{L} \in C^{0,\alpha}(\partial S)$.*

Then (R^-) has the (unique) solution

$$u = -\mathcal{V}^-(\sigma\varphi) + \mathcal{W}^-\varphi + q\varphi - \mathcal{V}^-\mathcal{L}.$$

Proof. (i) Let φ_0 be a solution of the homogeneous equation $(\mathcal{R}_D^+)$, that is,

$$V_0(\sigma\varphi_0) + (W_0 + \tfrac{1}{2}I)\varphi_0 = 0. \tag{1.82}$$

Then, by $(1.17)_2$, the function

$$\mathcal{U}^- = \mathcal{V}^-(\sigma\varphi_0) + \mathcal{W}^-\varphi_0 - p(\sigma\varphi_0)(\mathcal{V}^-\Phi) \tag{1.83}$$

satisfies

$$\Delta\mathcal{U}^- = 0,$$

$$\mathcal{U}^- = V_0(\sigma\varphi_0) + (W_0 + \tfrac{1}{2}I)\varphi_0 - p(\sigma\varphi_0)(V_0\Phi) = -\omega p(\sigma\varphi_0),$$

$$\mathcal{U}^-(x) = \bigl(p(\sigma\varphi_0)\bigr)\ln|x| - (p\Phi)\bigl(p(\sigma\varphi_0)\bigr)\ln|x| + \mathcal{U}^{\mathcal{A}}(x) = \mathcal{U}^{\mathcal{A}}(x) \quad \text{as } |x| \to \infty.$$

By Corollary 1.30(ii),

$$\omega p(\sigma\varphi_0) = 0, \quad \mathcal{U}^- = 0;$$

hence, since $\omega \neq 0$, it follows that $p(\sigma\varphi_0) = 0$ and, by (1.83),

$$\mathcal{V}^-(\sigma\varphi_0) + \mathcal{W}^-\varphi_0 = 0.$$

Applying ∂_ν to this equality, we find that

$$(W_0^* - \tfrac{1}{2}I)(\sigma\varphi_0) + N_0\varphi_0 = 0. \tag{1.84}$$

Then, by (1.82) and (1.84), the function $\mathcal{U}^+ = \mathcal{V}^+(\sigma\varphi_0) + \mathcal{W}^+\varphi_0$ satisfies

$$\Delta\mathcal{U}^+ = 0,$$

$$\partial_\nu\mathcal{U}^+ + \sigma\mathcal{U}^+|_{\partial S}$$
$$= \bigl[(W_0^* + \tfrac{1}{2}I)(\sigma\varphi_0) + N_0\varphi_0\bigr] + \sigma\bigl[V_0(\sigma\varphi_0) + (W_0 - \tfrac{1}{2}I)\varphi_0\bigr] = \sigma\varphi_0 - \sigma\varphi_0 = 0.$$

The unique solution of this homogeneous interior Robin problem is

$$\mathcal{U}^+ = \mathcal{V}^+(\sigma\varphi_0) + \mathcal{W}^+\varphi_0 = 0.$$

Operating with ∂_ν in the above equality, we find that

$$(W_0^* + \tfrac{1}{2}I)(\sigma\varphi_0) + N_0\varphi_0 = 0. \tag{1.85}$$

Next, we subtract (1.84) from (1.85) to obtain $\sigma\varphi_0 = 0$, which yields $\varphi_0 = 0$. Hence, since the homogeneous equation $(\mathcal{R}_D^+)$ has only the zero solution, from the

Fredholm Alternative we conclude that $(\mathcal{R}_D^+)$ has a unique solution $\varphi \in C^{0,\alpha}(\partial S)$. Using Theorem 1.34, we see that $V_0 \mathcal{K}$, $V_0(\sigma\varphi)$, $W_0\varphi \in C^{1,\alpha}(\partial S)$, so $\varphi \in C^{1,\alpha}(\partial S)$.

To verify that this solution also satisfies (1.80), we apply q to $(\mathcal{R}_D^+)$. According to Lemma 1.44(i),(iii),

$$\omega p(\sigma\varphi) - \tfrac{1}{2}q\varphi + \tfrac{1}{2}q\varphi = \omega(p\mathcal{K}),$$

or

$$\omega p(\sigma\varphi - \mathcal{K}) = 0,$$

which implies that (1.80) holds.

We need to check that the function $u = -\mathcal{V}^+(\sigma\varphi) - \mathcal{W}^+\varphi + \mathcal{V}^+\mathcal{K}$ is indeed the (unique) solution of (R^+). Again by Theorem 1.34, $u \in C^{1,\alpha}(\bar{S}^+)$ and $\Delta u = 0$ (in S^+). From the expression of u we see that

$$\begin{aligned}Tu + \sigma u|_{\partial S} &= (W_0^* + \tfrac{1}{2}I)(\mathcal{K} - \sigma\varphi) - N_0\varphi + \sigma\big[V_0(\mathcal{K} - \sigma\varphi) - (W_0 - \tfrac{1}{2}I)\varphi\big] \\ &= (W_0^* + \tfrac{1}{2}I)(\mathcal{K} - \sigma\varphi) - N_0\varphi + \sigma\varphi.\end{aligned}$$

To show that the right-hand side above is equal to $\mathcal{K}$, we consider the function

$$\mathcal{U} = \big[(W_0^* + \tfrac{1}{2}I)(\mathcal{K} - \sigma\varphi) - N_0\varphi + \sigma\varphi\big] - \mathcal{K} = (W_0^* - \tfrac{1}{2}I)(\mathcal{K} - \sigma\varphi) - N_0\varphi.$$

Using the composition formulae $(1.23)_1$ and $(1.24)_2$, we find that $\mathcal{U}$ satisfies

$$\begin{aligned}V_0\mathcal{U} &= \big(V_0(W_0^* - \tfrac{1}{2}I)\big)(\mathcal{K} - \sigma\varphi) - (V_0 N_0)\varphi \\ &= (W_0 - \tfrac{1}{2}I)V_0(\mathcal{K} - \sigma\varphi) - (W_0^2 - \tfrac{1}{4}I)\varphi \\ &= (W_0 - \tfrac{1}{2}I)\big[V_0(\mathcal{K} - \sigma\varphi) - (W_0 + \tfrac{1}{2}I)\varphi\big] = 0.\end{aligned}$$

Since $\omega \neq 0$, Theorem 1.43 implies that $\mathcal{U} = 0$, which yields the required boundary condition.

(ii) Applying q to $(\mathcal{R}_D^-)$ with $c = q\varphi$ and using Lemma 1.44(i),(iii), we find that

$$\omega p(\sigma\varphi) + \tfrac{1}{2}q\varphi + \tfrac{1}{2}q\varphi - q\varphi = -\omega(p\mathcal{L}),$$

which can be written as

$$\omega p(\sigma\varphi + \mathcal{L}) = 0.$$

Since $\omega \neq 0$, this means that any solution of $(\mathcal{R}_D^-)$ also satisfies (1.81).

Let φ_0 be a solution of the homogeneous equation $(\mathcal{R}_D^-)$ with $c = q\varphi$, that is,

$$V_0(\sigma\varphi_0) - (W_0 - \tfrac{1}{2}I)\varphi_0 - q\varphi_0 = 0. \tag{1.86}$$

Then the function $\mathcal{V}^+(\sigma\varphi_0) - \mathcal{W}^+\varphi_0 - q\varphi_0$ is the unique solution of the homogeneous problem (D$^+$), so
$$\mathcal{V}^+(\sigma\varphi_0) - \mathcal{W}^+\varphi_0 - q\varphi_0 = 0.$$

Applying ∂_ν to this equality yields
$$(W_0^* + \tfrac{1}{2}I)(\sigma\varphi_0) - N_0\varphi_0 = 0. \tag{1.87}$$

Since $\mathcal{L} = 0$, condition (1.81) reduces to $p\varphi_0 = 0$, so, by (1.86) and (1.87), the function
$$\mathcal{U}^- = \mathcal{V}^-(\sigma\varphi_0) - \mathcal{W}^-\varphi_0 - q\varphi_0$$
satisfies
$$\Delta\mathcal{U}^- = 0,$$
$$\partial_\nu\mathcal{U}^- - \sigma\mathcal{U}^-|_{\partial S} = \left[(W_0^* - \tfrac{1}{2}I)(\sigma\varphi_0) - N_0\varphi_0\right] - \sigma\left[V_0(\sigma\varphi_0) - (W_0 + \tfrac{1}{2}I)\varphi_0 - q\varphi_0\right]$$
$$= -\sigma\varphi_0 - \sigma(-\varphi_0) = 0,$$
$$\mathcal{U}^-(x) = \left(p(\sigma\varphi_0)\right)\ln|x| + \mathcal{U}^\mathcal{A}(x) - q\varphi_0 = \mathcal{U}^\mathcal{A}(x) - q\varphi_0 \quad \text{as } |x| \to \infty.$$

By Corollary 1.30(i),
$$q\varphi_0 = 0, \quad \mathcal{U}^- = \mathcal{V}^-(\sigma\varphi_0) - \mathcal{W}^-\varphi_0 - q\varphi_0 = \mathcal{V}^-(\sigma\varphi_0) - \mathcal{W}^-\varphi_0 = 0.$$

Operating with ∂_ν in the last equality leads to
$$(W_0^* - \tfrac{1}{2}I)(\sigma\varphi_0) - N_0\varphi_0 = 0. \tag{1.88}$$

We now subtract (1.88) from (1.87) and, as in (i), arrive at $\varphi_0 = 0$. Consequently, ($\mathcal{R}_D^-$) has a unique solution $\varphi \in C^{0,\alpha}(\partial S)$. We have seen that this solution satisfies condition (1.81). Also, by the mapping properties of the boundary operators, it belongs to $C^{1,\alpha}(\partial S)$.

Arguments similar to those used in (i) but based on ($\mathcal{R}_D^-$) with $c = q\varphi$ show that
$$u = -\mathcal{V}^-(\sigma\varphi) + \mathcal{W}^-\varphi + q\varphi - \mathcal{V}^-\mathcal{L} \in C^{1,\alpha}(\bar{S}^-)$$
is the (unique) solution of (R$^-$), which, because of (1.81), belongs to $\mathcal{A}^*$, as stipulated in Definition 1.21.

1.60. Remarks. (i) When $\omega = 0$, we need to use a modified fundamental solution of the form $g(x,y) + c$, where c is a suitably chosen constant, as we did for the Dirichlet problems in §1.9.

(ii) If $\sigma = \text{const}$, then Theorem 1.59 remains valid even when $\omega = 0$.

1.12. The substitute direct method

To avoid equations of the first kind in the Dirichlet problems, we alter the approach somewhat. More precisely, instead of restricting Green's representation formulae to the boundary we take the normal derivative of both sides of $(1.19)_1$ and (1.63) on ∂S and, in view of Theorem 1.34, find that

$$\partial_\nu u = (W_0^* + \tfrac{1}{2}I)(\partial_\nu u) - N_0(u|_{\partial S}),$$

$$\partial_\nu u = -(W_0^* - \tfrac{1}{2}I)(\partial_\nu u) + N_0(u|_{\partial S}),$$

which for $(D^\pm)$ leads to the "substitute" integral equations of the second kind

$$(W_0^* - \tfrac{1}{2}I)\varphi = N_0 \mathcal{P}, \qquad (\mathcal{D}_S^+)$$

$$(W_0^* + \tfrac{1}{2}I)\varphi = N_0 \mathcal{R}. \qquad (\mathcal{D}_S^-)$$

Clearly, $(\mathcal{D}_S^+)$, $(\mathcal{N}_D^-)$ and $(\mathcal{D}_S^-)$, $(\mathcal{N}_D^+)$ (see §1.11) are mutually adjoint. We must remember that φ still needs to satisfy condition (1.64), that is,

$$p\varphi = 0. \tag{1.89}$$

1.61. Theorem. (i) *The pair of equations $(\mathcal{D}_S^+)$ and (1.64) has a unique solution $\varphi \in C^{0,\alpha}(\partial S)$ for any $\mathcal{P} \in C^{1,\alpha}(\partial S)$, $\alpha \in (0,1)$.*

Then (D^+) has the (unique) solution

$$u = \mathcal{V}^+ \varphi - \mathcal{W}^+ \mathcal{P}. \tag{1.90}$$

(ii) *The pair of equations $(\mathcal{D}_S^-)$ and (1.64) has a unique solution $\varphi \in C^{0,\alpha}(\partial S)$ for any $\mathcal{R} \in C^{1,\alpha}(\partial S)$, $\alpha \in (0,1)$.*

Then (D^-) has the (unique) solution (in $\mathcal{A}^*$)

$$u = -\mathcal{V}^- \varphi + \mathcal{W}^- \mathcal{R} + q\mathcal{R}. \tag{1.91}$$

(iii) *$(\mathcal{N}_D^+)$ is solvable in $C^{1,\alpha}(\partial S)$ for any $\mathcal{Q} \in C^{0,\alpha}(\partial S)$ such that $p\mathcal{Q} = 0$, and its solution is unique up to an arbitrary constant.*

Then (N^+) has the family of solutions

$$u = \mathcal{V}^+ \mathcal{Q} - \mathcal{W}^+ \psi + c, \tag{1.92}$$

where ψ is any solution of $(\mathcal{N}_D^+)$ and c is an arbitrary constant.

(iv) $(\mathcal{N}_D^-)$ has a unique solution $\psi \in C^{1,\alpha}(\partial S)$ for any $\mathcal{S} \in C^{0,\alpha}(\partial S)$. Then, if $p\mathcal{S} = 0$, (N^-) has the (unique) solution (in $\mathcal{A}$)

$$u = -\mathcal{V}^- \mathcal{S} + \mathcal{W}^- \psi. \tag{1.93}$$

Proof. (i), (iv) The integral operators in $(\mathcal{D}_S^+)$ and $(\mathcal{N}_D^-)$ are the same as those in $(\mathcal{N}_C^-)$ and $(\mathcal{D}_C^+)$, respectively. Hence, by the arguments used in the proof of Theorem 1.45(i),(iv), $(\mathcal{D}_S^+)$ and $(\mathcal{N}_D^-)$ are uniquely solvable in $C^{0,\alpha}(\partial S)$. Also, applying p in $(\mathcal{D}_S^+)$ yields (1.89).

In the case of $(\mathcal{D}_S^+)$, $\varphi \in C^{0,\alpha}(\partial S)$ and $\mathcal{P} \in C^{1,\alpha}(\partial S)$ imply that u given by (1.90) belongs to $C^{1,\alpha}(\bar{S}^+)$. Also, $\Delta u = 0$ (in S^+). From (1.90) we now see that

$$u|_{\partial S} = V_0 \varphi - (W_0 - \tfrac{1}{2}I)\mathcal{P}. \tag{1.94}$$

The function

$$\mathcal{G} = \left[V_0\varphi - (W_0 - \tfrac{1}{2}I)\mathcal{P}\right] - \mathcal{P} = V_0\varphi - (W_0 + \tfrac{1}{2}I)\mathcal{P} \tag{1.95}$$

satisfies

$$\begin{aligned}N_0\mathcal{G} &= (N_0 V_0)\varphi - \bigl(N_0(W_0 + \tfrac{1}{2}I)\bigr)\mathcal{P} \\ &= (W_0^{*2} - \tfrac{1}{4}I)\varphi - (W_0^* + \tfrac{1}{2}I)(N_0\mathcal{P}) \\ &= (W_0^* + \tfrac{1}{2}I)\bigl[(W_0^* - \tfrac{1}{2}I)\varphi - N_0\mathcal{P}\bigr] = 0;\end{aligned}$$

therefore, by Theorem 1.39(ii), $\mathcal{G} = c = \text{const}$. From this, (1.95), (1.89) and Lemma 1.44(i),(iii) we find that

$$q\mathcal{G} = c = q(V_0\varphi) - q(W_0 + \tfrac{1}{2}I)\mathcal{P} = \omega(p\varphi) + \tfrac{1}{2}q\mathcal{P} - \tfrac{1}{2}q\mathcal{P} = 0,$$

which means that $\mathcal{G} = 0$. In turn, (1.94) shows that this is equivalent to $u|_{\partial S} = \mathcal{P}$. We have thus verified that (1.90) is the (unique) solution of (D^+).

In the case of $(\mathcal{N}_D^-)$, we have $\psi = 2(W_0\psi - V_0\mathcal{S}) \in C^{1,\alpha}(\partial S)$, so u defined by (1.93) belongs to $C^{1,\alpha}(\bar{S}^-)$. In addition, $\Delta u = 0$ (in S^-) and, if $p\mathcal{S} = 0$, then $u \in \mathcal{A}$. To show that u is the (unique) solution of (N^-), we also need to check the boundary condition for (1.93). That equality yields

$$Tu = -(W_0^* - \tfrac{1}{2}I)\mathcal{S} + N_0\psi, \tag{1.96}$$

and we operate with the function

$$\mathcal{H} = \left[-(W_0^* - \tfrac{1}{2}I)\mathcal{S} + N_0\psi\right] - \mathcal{S} = N_0\psi - (W_0^* + \tfrac{1}{2}I)\mathcal{S}, \tag{1.97}$$

which satisfies

$$V_0 \mathcal{H} = (V_0 N_0)\psi - \left(V_0(W_0^* + \tfrac{1}{2}I)\right)\mathcal{S}$$
$$= (W_0^2 - \tfrac{1}{4}I)\psi - (W_0 + \tfrac{1}{2}I)(V_0\mathcal{S})$$
$$= (W_0 + \tfrac{1}{2}I)\big[(W_0 - \tfrac{1}{2}I)\psi - V_0\mathcal{S}\big] = 0.$$

By Theorem 1.43, if $\omega \neq 0$, then $\mathcal{H} = 0$. If $\omega = 0$, then there is $c = \text{const}$ such that $\mathcal{H} = c\Phi$. Applying p to this and (1.97) and using Lemma 1.44(ii),(iv), we see that

$$c = c(p\Phi) = p\mathcal{H} = \tfrac{1}{2}p\mathcal{S} - \tfrac{1}{2}p\mathcal{S} = 0,$$

so $\mathcal{H} = 0$ in either case. By (1.96), this is equivalent to $Tu = \mathcal{S}$.

(iii) The null space of $W_0^* + \tfrac{1}{2}I$ is spanned by Φ, and, by Lemma 1.44(i),

$$(\Phi, V_0\mathcal{Q}) = q(V_0\mathcal{Q}) = \omega(p\mathcal{Q}) = 0;$$

hence, by the Fredholm Alternative, $(\mathcal{N}_D^+)$ is solvable in $C^{0,\alpha}(\partial S)$. Since the null space of $W_0 + \tfrac{1}{2}I$ is spanned by 1, the solution ψ is unique up to an arbitrary constant.

Just as in the case of $(\mathcal{N}_D^-)$, we have $\psi \in C^{1,\alpha}(\partial S)$, so u given by (1.92) belongs to $C^{1,\alpha}(\bar{S}^+)$. Also, $\Delta u = 0$ (in S^+). To check that u satisfies the boundary condition of (N$^+$), we apply T to (1.92), that is,

$$Tu = (W_0^* + \tfrac{1}{2}I)\mathcal{Q} - N_0\psi,$$

and convince ourselves that the function

$$\mathcal{X} = \big[(W_0^* + \tfrac{1}{2}I)\mathcal{Q} - N_0\psi\big] - \mathcal{Q} = (W_0^* - \tfrac{1}{2}I)\mathcal{Q} - N_0\psi$$

satisfies

$$V_0\mathcal{X} = \left(V_0(W_0^* - \tfrac{1}{2}I)\right)\mathcal{Q} - (V_0 N_0)\psi$$
$$= (W_0 - \tfrac{1}{2}I)(V_0\mathcal{Q}) - (W_0^2 - \tfrac{1}{4}I)\psi$$
$$= (W_0 - \tfrac{1}{2}I)\big[V_0\mathcal{Q} - (W_0 + \tfrac{1}{2}I)\psi\big] = 0.$$

As in (iv), we use this in conjunction with the equality

$$p\mathcal{X} = p(W_0^*\mathcal{Q}) - \tfrac{1}{2}p\mathcal{Q} - p(N_0\psi) = -p\mathcal{Q} = 0$$

to conclude that $\mathcal{X} = 0$, which is the same as $Tu = \mathcal{Q}$.

(ii) By Lemma 1.44(iv),
$$(1, N_0\mathcal{R}) = p(N_0\mathcal{R}) = 0,$$
so, by the Fredholm Alternative, $(\mathcal{D}_S^-)$ is solvable in $C^{0,\alpha}(\partial S)$ and its solutions are
$$\varphi = \varphi_0 + a\Phi, \tag{1.98}$$
where φ_0 is any (fixed) solution and a is an arbitrary constant. However, we are interested only in the function (1.98) that also satisfies (1.89), and this one clearly corresponds to $a = -p\varphi_0$; that is,
$$\varphi = \varphi_0 - (p\varphi_0)\Phi.$$
The difference $\bar{\varphi}$ of any two such solutions satisfies
$$(W_0^* + \tfrac{1}{2}I)\bar{\varphi} = 0, \quad p\bar{\varphi} = 0,$$
so $\bar{\varphi} = a\Phi$, $a = $ const. But then $p\bar{\varphi} = a(p\Phi) = a = 0$, which means that $\bar{\varphi} = 0$. Thus, we conclude that the pair of equations $(\mathcal{D}_S^-)$ and (1.89) has a unique solution, independent of the solution φ_0 of $(\mathcal{D}_S^-)$ used in its construction.

To check that u given by (1.91) is the (unique) solution of (D^-), we note first that $u \in C^{1,\alpha}(\bar{S}^-)$, $Au = 0$ (in S^-), and $u \in \mathcal{A}^*$ (because of (1.89)). As regards the boundary condition, (1.91) yields
$$u|_{\partial S} = -V_0\varphi + (W_0 + \tfrac{1}{2}I)\mathcal{R} + q\mathcal{R},$$
and we easily see that the function
$$\mathcal{Y} = \left[-V_0\varphi + (W_0 + \tfrac{1}{2}I)\mathcal{R} + q\mathcal{R}\right] - \mathcal{R} = -V_0\varphi + (W_0 - \tfrac{1}{2}I)\mathcal{R} + q\mathcal{R}$$
satisfies
$$\begin{aligned}
N_0\mathcal{Y} &= -(N_0V_0)\varphi + \left(N_0(W_0 - \tfrac{1}{2}I)\right)\mathcal{R} \\
&= -(W_0^{*2} - \tfrac{1}{4}I)\varphi + (W_0^* - \tfrac{1}{2}I)(N_0\mathcal{R}) \\
&= -(W_0^* - \tfrac{1}{2}I)\left[(W_0^* + \tfrac{1}{2}I)\varphi - N_0\mathcal{R}\right] = 0
\end{aligned}$$
and
$$q\mathcal{Y} = -q(V_0\varphi) + q(W_0\mathcal{R}) - \tfrac{1}{2}q\mathcal{R} + q\mathcal{R} = -\mathcal{C}(p\varphi) - \tfrac{1}{2}q\mathcal{R} - \tfrac{1}{2}q\mathcal{R} + q\mathcal{R} = 0,$$
which, as in (i), leads to $u|_{\partial S} = \mathcal{R}$.

1.62. Remark. The operators of $(\mathcal{D}_C^+)$, $(\mathcal{D}_C^-)$ are adjoint to those of $(\mathcal{D}_S^+)$, $(\mathcal{D}_S^-)$, and the operators of $(\mathcal{N}_C^+)$, $(\mathcal{N}_C^-)$ are adjoint to those of $(\mathcal{N}_D^+)$, $(\mathcal{N}_D^-)$.

1.63. Remark. There is a substitute method for the Robin problems as well. Instead of $u|_{\partial S}$, we can take $\partial_\nu u$ as the unknown function in the boundary integral equations for $(\mathrm{R}^\pm)$.

Restricting the representation formula $(1.19)_1$ to ∂S, we find that

$$V_0(\partial_\nu u) - (W_0 + \tfrac{1}{2}I)(u|_{\partial S}) = 0. \tag{1.99}$$

If we set

$$\theta = (\partial_\nu u)/\sigma, \tag{1.100}$$

then the boundary condition for (R^+) can be written as $u|_{\partial S} = -\theta + \mathcal{K}/\sigma$ and (1.99) reduces to

$$V_0(\sigma\theta) + (W_0 + \tfrac{1}{2}I)\theta = (W_0 + \tfrac{1}{2}I)(\mathcal{K}/\sigma). \tag{$\mathcal{R}_S^+$}$$

Proceeding in the same way with the representation (1.63), we arrive at

$$V_0(\partial_\nu u) - (W_0 - \tfrac{1}{2}I)(u|_{\partial S}) - c = 0,$$

and the substitution $u|_{\partial S} = \theta - \mathcal{L}/\sigma$, with θ as in (1.100), leads to the boundary integral equation

$$V_0(\sigma\theta) - (W_0 - \tfrac{1}{2}I)\theta - c = -(W_0 - \tfrac{1}{2}I)(\mathcal{L}/\sigma). \tag{$\mathcal{R}_S^-$}$$

Since $\sigma\theta = \partial_\nu u$, we must remember to impose on θ, in both $(\mathcal{R}_S^+)$ and $(\mathcal{R}_S^-)$, the additional condition

$$p(\sigma\theta) = 0. \tag{1.101}$$

Noting that the operators on the left-hand side in $(\mathcal{R}_S^+)$ and $(\mathcal{R}_S^-)$ coincide with those in $(\mathcal{R}_D^+)$ and $(\mathcal{R}_D^-)$, we conclude that the substitute equations for the Robin problems are manipulated in the same way as those in the direct method.

It is easy to see that the (unique) solution of $(\mathcal{R}_S^+)$ satisfies (1.101). Applying q to the former, we obtain $\omega p(\sigma\theta) = 0$, which, if $\omega \neq 0$ (as stipulated in the solvability theorem), yields $p(\sigma\theta) = 0$.

The same technique used in $(\mathcal{R}_S^-)$ leads to

$$c = \omega p(\sigma\theta) + q(\theta - \mathcal{L}/\sigma).$$

THE SUBSTITUTE DIRECT METHOD 53

In this case, the procedure developed in the direct method indicates that $(\mathcal{R}_S^-)$ with $c = q(\theta - \mathcal{L}/\sigma)$ has a unique solution. Consequently, if $\omega \neq 0$, then here we also find that $p(\sigma\theta) = 0$.

As in §1.11, it is not difficult to verify that the functions suggested by the representation formulae, namely,

$$u = \mathcal{V}^+(\sigma\theta) - \mathcal{W}^+(-\theta + \mathcal{K}/\sigma),$$
$$u = -\mathcal{V}^-(\sigma\theta) + \mathcal{W}^+(\theta - \mathcal{L}/\sigma) + q(\theta - \mathcal{L}/\sigma),$$

are, respectively, the (unique) solutions of (R^+) and (R^-) when $\omega \neq 0$.

Chapter 2
Plane Strain

2.1. Notation and prerequisites

In what follows $\mathcal{M}_{p\times q}$ is the vector space of $(p \times q)$-matrices; $H^{(i)}$, $i = 1, \ldots, q$, are the columns of a matrix $H \in \mathcal{M}_{p\times q}$; H^{T} is the transpose of H; $\{E_{ij}\}_{i,j=1}^{n}$ is the standard ordered basis for the subspace of constant matrices in $\mathcal{M}_{n\times n}$; and $E_n = E_{11} + \cdots + E_{nn}$ is the identity $(n \times n)$-matrix.

If X is a space of scalar functions and $\varphi \in \mathcal{M}_{p\times q}$, then $\varphi \in X$ means that every component of φ belongs to X. For simplicity, we use the term 'function' in both the scalar and matrix-valued cases.

Next, if $\Theta \in \mathcal{M}_{p\times q}$ and L is an operator defined on functions $\theta \in \mathcal{M}_{p\times 1}$ and such that $L\theta \in \mathcal{M}_{r\times 1}$, then $L\Theta \in \mathcal{M}_{r\times q}$ is the matrix with columns $(L\Theta)^{(i)} = L\Theta^{(i)}$, $i = 1, \ldots, q$. Also, if v is an operator defined on scalar functions, then $v\theta$ is the $(p \times 1)$-vector of components $v\theta_j$, $j = 1, \ldots, p$.

From now on, unless otherwise stipulated, Greek and Latin indices take the values $1, 2$ and $1, 2, 3$, respectively, the convention of summation over repeated indices is understood, and we write
$$\partial_i \psi = \psi_{,i} = \partial \psi / \partial x_i.$$

The equilibrium equations of three-dimensional elasticity are
$$t_{ij,j} + f_i = 0, \tag{2.1}$$

where $t_{ij} = t_{ji}$ and f_i are the components of the stress tensor and body force vector, respectively. These equations are used in conjunction with the constitutive relations, which for a homogeneous and isotropic material are
$$t_{ij} = \lambda u_{k,k} \delta_{ij} + \mu(u_{i,j} + u_{j,i}), \tag{2.2}$$

where u_i are the displacements and λ and μ are the elastic (Lamé) coefficients.

The components of the stress vector t in a direction $n = (n_1, n_2, n_3)^{\mathrm{T}}$ are
$$t_i = t_{ij} n_j, \tag{2.3}$$
and
$$\mathcal{E} = \tfrac{1}{4} t_{ij}(u_{i,j} + u_{j,i}) = \tfrac{1}{2} t_{ij} u_{i,j} \tag{2.4}$$

is the internal energy density.

NOTATION AND PREREQUISITES

In the theory of plane strain it is assumed that

$$u_\alpha = u_\alpha(x_1, x_2), \quad u_3 = 0, \tag{2.5}$$
$$f_\alpha = f_\alpha(x_1, x_2), \quad f_3 = 0,$$
$$n = (n_1, n_2, 0)^T.$$

Then (2.1)–(2.4) yield the equilibrium equations, constitutive relations, stress vector components and internal energy density, respectively, in the form

$$t_{\alpha\beta,\beta} + f_\alpha = 0, \tag{2.6}$$
$$t_{\alpha\beta} = \lambda u_{\gamma,\gamma}\delta_{\alpha\beta} + \mu(u_{\alpha,\beta} + u_{\beta,\alpha}), \tag{2.7}$$
$$t_\alpha = t_{\alpha\beta}n_\beta, \tag{2.8}$$
$$\mathcal{E} = \tfrac{1}{4}t_{\alpha\beta}(u_{\alpha,\beta} + u_{\beta,\alpha}) = \tfrac{1}{2}t_{\alpha\beta}u_{\alpha,\beta}. \tag{2.9}$$

Replacing (2.7) in (2.6), we obtain the equilibrium equations in terms of the displacements; that is,

$$(\lambda + \mu)u_{\beta,\beta\alpha} + \mu u_{\alpha,\beta\beta} + f_\alpha = 0,$$

or

$$(\lambda + \mu)\,\mathrm{grad}\,\mathrm{div}\,u + \mu\Delta u + f = 0, \tag{2.10}$$

where $u = (u_1, u_2)^T$, $f = (f_1, f_2)^T$, and $\Delta = \partial_1^2 + \partial_2^2$ is the two-dimensional Laplacian. In the absence of body forces ($f = 0$), (2.10) can be written as

$$A(\partial_1, \partial_2)u = 0, \tag{2.11}$$

where $A(\partial_1, \partial_2)$ is the matrix differential operator defined by

$$A(\partial_1, \partial_2) = \begin{pmatrix} \mu\Delta + (\lambda + \mu)\partial_1^2 & (\lambda + \mu)\partial_1\partial_2 \\ (\lambda + \mu)\partial_1\partial_2 & \mu\Delta + (\lambda + \mu)\partial_2^2 \end{pmatrix}. \tag{2.12}$$

Throughout this chapter we assume that

$$\lambda + \mu > 0, \quad \mu > 0. \tag{2.13}$$

These conditions are less restrictive than those imposed on the Lamé coefficients in three-dimensional elasticity, namely,

$$3\lambda + 2\mu > 0, \quad \mu > 0.$$

Clearly, if the latter are satisfied, then so are inequalities (2.13), but the converse is not necessarily true.

2.1. Theorem. *System (2.11) is elliptic.*

Proof. From (2.12) it follows that

$$(\det A)(\partial_1, \partial_2) = \mu(\lambda + 2\mu)\Delta\Delta, \tag{2.14}$$

and conditions (2.13) imply that $\mu(\lambda + 2\mu) \neq 0$; therefore, the matrix $A(\xi_1, \xi_2)$ is invertible for all $(\xi_1, \xi_2) \neq 0$, which means that (2.11) is an elliptic system.

Suppose that S^+ or S^- (defined as in Chapter 1) is occupied by a homogeneous and isotropic material. Using formulae (2.7) in (2.8) (with $n_\alpha = \nu_\alpha$), we find that the components of the stress vector on the boundary ∂S can be written in the form

$$t_\alpha = t_{\alpha\beta}\nu_\beta = T_{\alpha\beta}u_\beta, \tag{2.15}$$

where the boundary stress operator T is defined by

$$T(\partial_1, \partial_2) = \begin{pmatrix} (\lambda + 2\mu)\nu_1\partial_1 + \mu\nu_2\partial_2 & \mu\nu_2\partial_1 + \lambda\nu_1\partial_2 \\ \lambda\nu_2\partial_1 + \mu\nu_1\partial_2 & \mu\nu_1\partial_1 + (\lambda + 2\mu)\nu_2\partial_2 \end{pmatrix}. \tag{2.16}$$

Finally, substituting (2.7) in (2.9) leads to

$$\mathcal{E} = \mathcal{E}(u,u) = \tfrac{1}{2}\big[(\lambda + 2\mu)(u_{1,1}^2 + u_{2,2}^2) + 2\lambda u_{1,1}u_{2,2} + \mu(u_{1,2} + u_{2,1})^2\big].$$

A general three-dimensional rigid displacement is of the form

$$(a_1 + b_3 x_2 - b_2 x_3,\ a_2 + b_1 x_3 - b_3 x_1,\ a_3 + b_2 x_1 - b_1 x_2)^{\mathrm{T}},$$

where a_i and b_i are constants. Hence, comparing with (2.5), we deduce that the general form of a two-dimensional rigid displacement is

$$(a_1 + b_3 x_2,\ a_2 - b_3 x_1)^{\mathrm{T}}. \tag{2.17}$$

Let $\mathcal{F}$ be the vector space of all such displacements, and let

$$F = \begin{pmatrix} 1 & 0 & x_2 \\ 0 & 1 & -x_1 \end{pmatrix}. \tag{2.18}$$

The columns $F^{(i)}$ of F form a basis for $\mathcal{F}$, and, using (2.12), (2.16) and (2.18), we can easily check that
$$(AF)(x) = 0, \quad x \in \mathbb{R}^2,$$
$$(TF)(x) = 0, \quad x \in \partial S.$$

Also, if we set
$$k_1 = a_1, \quad k_2 = a_2, \quad k_3 = b_3,$$
then we can write (2.17) as Fk, with a constant $k \in \mathcal{M}_{3\times 1}$.

2.2. Theorem. $\mathcal{E}(u,u)$ is a positive quadratic form, and $\mathcal{E}(u,u) = 0$ if and only if $u \in \mathcal{F}$.

Proof. We have
$$\mathcal{E}(u,u) = \tfrac{1}{2}\big[(\lambda + 2\mu)u_{1,1}^2 + 2\lambda u_{1,1}u_{2,2} + (\lambda + 2\mu)u_{2,2}^2 + \mu(u_{1,2} + u_{2,1})^2\big],$$
and the quadratic form
$$(\lambda + 2\mu)u_{1,1}^2 + 2\lambda u_{1,1}u_{2,2} + (\lambda + 2\mu)u_{2,2}^2$$
is positive if
$$\lambda^2 - (\lambda + 2\mu)^2 = -4\mu(\lambda + \mu) < 0,$$
which is guaranteed by (2.13). Clearly, $\mathcal{E}(u,u) = 0$ if and only if
$$u_{1,1} = u_{2,2} = u_{1,2} + u_{2,1} = 0,$$
which is equivalent to (2.17).

2.3. Lemma. If $u \in C^2(S^+) \cap C^1(\bar{S}^+)$, then
$$\int_{S^+} F^{\mathrm{T}}(Au)\,da = \int_{\partial S} F^{\mathrm{T}}(Tu)\,ds. \tag{2.19}$$

Proof. Since (2.11) is equivalent to (2.6) with $f_\alpha = 0$, it follows that, by the divergence theorem and (2.15),
$$\int_{S^+} (Au)_\alpha\,da = \int_{S^+} A_{\alpha\beta} u_\beta\,da = \int_{S^+} t_{\alpha\beta,\beta}\,da$$
$$= \int_{\partial S} t_{\alpha\beta}\nu_\beta\,ds = \int_{\partial S} T_{\alpha\beta} u_\beta\,ds = \int_{\partial S} (Tu)_\alpha\,ds, \tag{2.20}$$

and

$$\int_{S^+} (x_2 A_{1\beta} - x_1 A_{2\beta}) u_\beta \, da$$

$$= \int_{S^+} (x_2 t_{1\beta,\beta} - x_1 t_{2\beta,\beta}) \, da = \int_{S^+} \left[(x_2 t_{1\beta} - x_1 t_{2\beta})_{,\beta} - t_{1\beta}\delta_{2\beta} + t_{2\beta}\delta_{1\beta} \right] da$$

$$= \int_{\partial S} (x_2 t_{1\beta}\nu_\beta - x_1 t_{2\beta}\nu_\beta) \, ds = \int_{\partial S} (x_2 T_{1\beta} - x_1 T_{2\beta}) u_\beta \, ds. \qquad (2.21)$$

Equalities (2.20) and (2.21) can be written together as (2.19).

2.2. The fundamental boundary value problems

Let $\mathcal{A}$ be the vector space of functions $u \in \mathcal{M}_{2\times 1}$ in S^- which, in terms of polar coordinates r, θ with the pole at the origin, as $r = |x| \to \infty$ admit an asymptotic expansion of the form

$$\begin{aligned}
u_1(r,\theta) &= r^{-1}(\gamma m_0 \sin\theta + m_1 \cos\theta + m_0 \sin 3\theta + m_2 \cos 3\theta) + \tilde{u}_1(r,\theta), \\
u_2(r,\theta) &= r^{-1}(m_3 \sin\theta + \gamma m_0 \cos\theta + m_4 \sin 3\theta - m_0 \cos 3\theta) + \tilde{u}_2(r,\theta),
\end{aligned} \qquad (2.22)$$

where $m_0, \ldots, m_4$ are arbitrary constants,

$$\gamma = (\lambda + 3\mu)(\lambda + \mu)^{-1} \qquad (2.23)$$

and

$$\tilde{u}_\alpha = O(r^{-2}), \quad \partial_r \tilde{u} = O(r^{-3}), \quad \partial_\theta \tilde{u} = O(r^{-2}), \qquad (2.24)$$

uniformly with respect to θ.

Also, we define

$$\mathcal{A}^* = \mathcal{A} \oplus \mathcal{F};$$

that is, a generic element of $\mathcal{A}^*$ is written as $u = u^{\mathcal{A}} + Fk$, with an arbitrary constant $k \in \mathcal{M}_{3\times 1}$.

Extending the convention adopted in Chapter 1, if u is a function defined in a domain that includes ∂S, then we denote its restriction to ∂S by $u|_{\partial S}$, but we simply write Tu instead of $(Tu)|_{\partial S}$, since T is itself a boundary operator.

2.4. Definition. Let $\mathcal{P}, \mathcal{Q}, \mathcal{K}, \mathcal{R}, \mathcal{S}, \mathcal{L} \in C(\partial S)$ be given (2×1)-matrix functions, and let $\sigma \in C(\partial S)$ be a given positive definite (2×2)-matrix function. We consider the following interior and exterior Dirichlet, Neumann and Robin problems.

(D$^+$) Find $u \in C^2(S^+) \cap C^1(\bar{S}^+)$ such that

$$(Au)(x) = 0, \quad x \in S^+, \quad \text{and} \quad u(x) = \mathcal{P}(x), \quad x \in \partial S.$$

(N$^+$) Find $u \in C^2(S^+) \cap C^1(\bar{S}^+)$ such that

$$(Au)(x) = 0, \quad x \in S^+, \quad \text{and} \quad (Tu)(x) = \mathcal{Q}(x), \quad x \in \partial S.$$

(R$^+$) Find $u \in C^2(S^+) \cap C^1(\bar{S}^+)$ such that

$$(Au)(x) = 0, \quad x \in S^+, \quad \text{and} \quad (Tu + \sigma u)(x) = \mathcal{K}(x), \quad x \in \partial S.$$

(D$^-$) Find $u \in C^2(S^-) \cap C^1(\bar{S}^-) \cap \mathcal{A}^*$ such that

$$(Au)(x) = 0, \quad x \in S^-, \quad \text{and} \quad u(x) = \mathcal{R}(x), \quad x \in \partial S.$$

(N$^-$) Find $u \in C^2(S^-) \cap C^1(\bar{S}^-) \cap \mathcal{A}$ such that

$$(Au)(x) = 0, \quad x \in S^-, \quad \text{and} \quad (Tu)(x) = \mathcal{S}(x), \quad x \in \partial S.$$

(R$^-$) Find $u \in C^2(S^-) \cap C^1(\bar{S}^-) \cap \mathcal{A}^*$ such that

$$(Au)(x) = 0, \quad x \in S^-, \quad \text{and} \quad (Tu - \sigma u)(x) = \mathcal{L}(x), \quad x \in \partial S.$$

A function satisfying any one of the above sets of equations is called a *regular solution* of that boundary value problem, or simply a *solution*.

2.5. Remark. The same boundary value problems for the non-homogeneous system (2.11) can always be reduced to the corresponding ones in Definition 2.4 by means of a particular solution of the non-homogeneous system, usually constructed in terms of a so-called area potential.

2.6. Theorem. *If* (N$^+$) *and* (N$^-$) *are solvable, then*

$$p\mathcal{Q} = 0, \quad p\mathcal{S} = 0,$$

respectively, where p is the operator defined on $C(\partial S)$ by

$$p\varphi = \int_{\partial S} F^T \varphi \, ds.$$

Proof. If u is a solution of (N^+), then, by (2.19),

$$0 = \int_{S^+} F^T(Au)\, da = p(Tu) = pQ.$$

Now let $u \in \mathcal{A}$ be a solution of (N^-), and consider a disk K_R with the centre at the origin (in S^+) and radius R sufficiently large so that $\bar{S}^+$ lies strictly inside K_R. Applying (2.19) in $\bar{S}^- \cap \bar{K}_R$, we find that

$$0 = \int_{S^- \cap K_R} F^T(Au)\, da = \left(-\int_{\partial S} + \int_{\partial K_R}\right) F^T(Tu)\, ds, \qquad (2.25)$$

where we have taken into account the fact that the outward normal to $S^- \cap K_R$ on ∂S is directed into S^+.

Since on the circle ∂K_R

$$x_1 = R\cos\theta, \quad x_2 = R\sin\theta,$$
$$\partial_1 = (\cos\theta)\partial_R - (\sin\theta)R^{-1}\partial_\theta, \quad \partial_2 = (\sin\theta)\partial_R + (\cos\theta)R^{-1}\partial_\theta, \qquad (2.26)$$
$$\nu_1 = \cos\theta, \quad \nu_2 = \sin\theta,$$

from (2.16) and (2.24) it follows that, for R large,

$$(Tu)(R,\theta) = O(R^{-2}), \quad (T\tilde{u})(R,\theta) = O(R^{-3}). \qquad (2.27)$$

The first of these inequalities implies that

$$\int_{\partial K_R} \left(F^T(Tu)\right)_\alpha ds = \int_{\partial K_R} F_{\beta\alpha}(Tu)_\beta\, ds = \int_{\partial K_R} \delta_{\beta\alpha}(Tu)_\beta\, ds$$
$$= \int_{\partial K_R} (Tu)_\alpha\, ds = \int_0^{2\pi} O(R^{-2}) R\, d\theta \to 0 \quad \text{as } R \to \infty. \qquad (2.28)$$

Also, direct computation by means of (2.22), (2.26) and $(2.27)_2$ shows that

$$\int_{\partial K_R} \left(F^T(Tu)\right)_3 ds = \int_{\partial K_R} F_{\beta 3}(Tu)_\beta\, ds = \int_{\partial K_R} [x_2(Tu)_1 - x_1(Tu)_2]\, ds$$
$$= \int_0^{2\pi} [(\sin\theta)(T_{1\alpha}u_\alpha) - (\cos\theta)(T_{2\alpha}u_\alpha)] R^2\, d\theta$$

$$= \int_0^{2\pi} \{(-\sin\theta)[(m_1 + m_3)\cos\theta + 4m_0 \sin 3\theta + (m_3 + 3m_4)\cos 3\theta]$$
$$+ (\cos\theta)[(m_1 + m_3)\sin\theta + (m_4 + 3m_2)\sin 3\theta - 4m_0 \cos 3\theta]\} d\theta$$
$$+ O(R^{-1}) \to 0 \quad \text{as } R \to \infty. \tag{2.29}$$

From (2.25), (2.28) and (2.29) we now conclude that $p\mathcal{S} = 0$.

2.3. The Betti and Somigliana formulae

The next assertion is the analogue of Green's first identity with $u = v$.

2.7. Theorem. If $u \in C^2(S^+) \cap C^1(\bar{S}^+)$, then

$$\int_{S^+} u^T(Au)\, da + 2\int_{S^+} \mathcal{E}(u,u)\, da = \int_{\partial S} u^T(Tu)\, ds. \tag{2.30}$$

Proof. Integrating by parts, we obtain

$$\int_{S^+} u^T(Au)\, da = \int_{S^+} u_\alpha t_{\alpha\beta,\beta}\, da = \int_{S^+} \left[(u_\alpha t_{\alpha\beta})_{,\beta} - u_{\alpha,\beta} t_{\alpha\beta}\right] da$$
$$= \int_{\partial S} u_\alpha t_{\alpha\beta} \nu_\beta\, ds - \int_{S^+} \tfrac{1}{2} t_{\alpha\beta}(u_{\alpha,\beta} + u_{\beta,\alpha})\, da$$
$$= \int_{\partial S} u^T(Tu)\, ds - 2\int_{S^+} \mathcal{E}(u,u)\, da,$$

which yields (2.30).

The so-called *Betti formulae* are an immediate consequence of Theorem 2.7.

2.8. Corollary. (i) If $u \in C^2(S^+) \cap C^1(\bar{S}^+)$ satisfies $(Au)(x) = 0$, $x \in S^+$, then

$$2\int_{S^+} \mathcal{E}(u,u)\, da = \int_{\partial S} u^T(Tu)\, ds. \tag{2.31}$$

(ii) If $u \in C^2(S^-) \cap C^1(\bar{S}^-) \cap \mathcal{A}^*$ satisfies $(Au)(x) = 0$, $x \in S^-$, then

$$2\int_{S^-} \mathcal{E}(u,u)\, da = -\int_{\partial S} u^T(Tu)\, ds. \tag{2.32}$$

Proof. (i) This follows directly from (2.30).

(ii) Let K_R be a disk with the centre at the origin and radius R sufficiently large so that $\bar{S}^+$ lies strictly inside K_R. Applying (2.31) in $\bar{S}^- \cap \bar{K}_R$ and noting that here the outward normal on ∂S is directed into S^+, we obtain

$$2 \int_{S^- \cap K_R} \mathcal{E}(u,u)\, da = \left(-\int_{\partial S} + \int_{\partial K_R} \right) u^{\mathrm{T}}(Tu)\, ds.$$

We need to investigate the asymptotic behaviour of the integral over ∂K_R as $R \to \infty$. Suppose first that $u = u^{\mathcal{A}} \in \mathcal{A}$. By (2.22) and (2.27),

$$\int_{\partial K_R} (u^{\mathcal{A}})^{\mathrm{T}} (Tu^{\mathcal{A}})\, ds = \int_0^{2\pi} O(R^{-1}) \cdot O(R^{-2}) \cdot R\, d\theta \to 0 \quad \text{as } R \to \infty.$$

Then, using (2.28) and (2.29), for $u = u^{\mathcal{A}} + Fk$ we find that

$$\int_{\partial K_R} (u^{\mathcal{A}} + Fk)^{\mathrm{T}}(Tu^{\mathcal{A}})\, ds \to 0 \quad \text{as } R \to \infty,$$

which leads to (2.32). ∎

The analogue of Green's second identity is called the *reciprocity relation*.

2.9. Theorem. *If $u, v \in C^2(S^+) \cap C^1(\bar{S}^+)$, then*

$$\int_{S^+} \left[u^{\mathrm{T}}(Av) - v^{\mathrm{T}}(Au) \right] da = \int_{\partial S} \left[u^{\mathrm{T}}(Tv) - v^{\mathrm{T}}(Tu) \right] ds. \tag{2.33}$$

Proof. Let $s_{\alpha\beta}$ be the stresses generated by the displacements v_α. Then

$$\int_{S^+} \left[u^{\mathrm{T}}(Av) - v^{\mathrm{T}}(Au) \right] ds = \int_{S^+} (u_\alpha s_{\alpha\beta,\beta} - v_\alpha t_{\alpha\beta,\beta})\, ds$$

$$= \int_{S^+} \left[(u_\alpha s_{\alpha\beta} - v_\alpha t_{\alpha\beta})_{,\beta} - (u_{\alpha,\beta} s_{\alpha\beta} - v_{\alpha,\beta} t_{\alpha\beta}) \right] da$$

$$= \int_{\partial S} (u_\alpha s_{\alpha\beta} - v_\alpha t_{\alpha\beta}) \nu_\beta - \int_{S^+} (u_{\alpha,\beta} s_{\alpha\beta} - v_{\alpha,\beta} t_{\alpha\beta})\, da.$$

Using (2.7), we see that

$$u_{\alpha,\beta} s_{\alpha\beta} - v_{\alpha,\beta} t_{\alpha\beta}$$
$$= \lambda u_{\alpha,\alpha} v_{\gamma,\gamma} + \mu u_{\alpha,\beta}(v_{\alpha,\beta} + v_{\beta,\alpha}) - \lambda v_{\alpha,\alpha} u_{\gamma,\gamma} - \mu v_{\alpha,\beta}(u_{\alpha,\beta} + u_{\beta,\alpha}) = 0,$$

which proves (2.33). ∎

THE BETTI AND SOMIGLIANA FORMULAE

At this stage we need to construct a matrix of fundamental solutions for the operator $-A$. If we replace u in (2.11) by

$$D(x,y) = (\operatorname{adj} A)(\partial_x)\bigl[t(x,y)E_2\bigr], \tag{2.34}$$

where the symbol ∂_x indicates that the operator acts with respect to the point x, then

$$A(\partial_x)D(x,y) = A(\partial_x)(\operatorname{adj} A)(\partial_x)\bigl[t(x,y)E_2\bigr] = -\delta(|x-y|)E_2$$

if t is a solution of the equation

$$(\det A)(\partial_x)t(x,y) = -\delta(|x-y|).$$

By (2.14), this means that

$$(\Delta\Delta)(\partial_x)t(x,y) = -\bigl[\mu(\lambda+2\mu)\bigr]^{-1}\delta(|x-y|). \tag{2.35}$$

Since $(8\pi)^{-1}|x-y|^2\ln|x-y|$ is a fundamental solution for $\Delta\Delta$, that is,

$$(\Delta\Delta)(\partial_x)(|x-y|^2\ln|x-y|) = 8\pi\delta(|x-y|),$$

from (2.35) it follows that

$$t(x,y) = -\bigl[8\pi\mu(\lambda+2\mu)\bigr]^{-1}|x-y|^2\ln|x-y|. \tag{2.36}$$

Therefore, a matrix $D(x,y)$ of fundamental solutions for $-A$ is given by (2.34) with t as in (2.36) and, from (2.12),

$$(\operatorname{adj} A)(\partial_1,\partial_2) = \begin{pmatrix} \mu\Delta + (\lambda+\mu)\partial_2^2 & -(\lambda+\mu)\partial_1\partial_2 \\ -(\lambda+\mu)\partial_1\partial_2 & \mu\Delta + (\lambda+\mu)\partial_1^2 \end{pmatrix}. \tag{2.37}$$

This immediately implies the symmetry property

$$D(x,y) = D(y,x) = D^{\mathrm{T}}(x,y). \tag{2.38}$$

Along with $D(x,y)$ we also consider the matrix

$$P(x,y) = [T(\partial_y)D(y,x)]^{\mathrm{T}}, \tag{2.39}$$

which plays an essential role in our arguments.

Using (2.36) and (2.37), from (2.34) and (2.39) we obtain the explicit expressions

$$D(x,y) = -\left[4\pi\mu(\gamma+1)\right]^{-1}\left[(2\gamma\ln|x-y|+2\gamma+1)E_2 \right.$$
$$\left. - 2(x_\alpha - y_\alpha)(x_\beta - y_\beta)|x-y|^{-2}E_{\alpha\beta}\right], \qquad (2.40)$$

$$P(x,y) = -(2\pi)^{-1}\left[(\gamma-1)(\gamma+1)^{-1}(\partial_{s(y)}\ln|x-y|)\varepsilon_{\alpha\beta}E_{\alpha\beta} \right.$$
$$+ (\partial_{\nu(y)}\ln|x-y|)E_2$$
$$\left. - 2(\gamma+1)^{-1}\partial_{s(y)}\bigl((x_\alpha - y_\alpha)(x_\beta - y_\beta)|x-y|^{-2}\bigr)\varepsilon_{\alpha\delta}E_{\delta\beta}\right], \quad (2.41)$$

where γ is given by (2.23) and $\varepsilon_{\alpha\beta}$ is the two-dimensional Ricci tensor defined by

$$\varepsilon_{\alpha\beta} = \begin{cases} 1 & \text{if } \alpha = 1,\ \beta = 2, \\ -1 & \text{if } \alpha = 2,\ \beta = 1, \\ 0 & \text{if } \alpha = \beta. \end{cases}$$

Since, as seen above, for all $x \in \mathbb{R}^2$, $x \neq y$, we have

$$A(\partial_x)D(x,y) = 0$$

and, by (2.38) and (2.39),

$$\left[A(\partial_x)P^{(i)}(x,y)\right]_j = A_{jk}(\partial_x)P_k^{(i)}(x,y) = A_{jk}(\partial_x)\left[T_{il}(\partial_y)D_{lk}(y,x)\right]$$
$$= T_{il}(\partial_y)\left[A_{jk}(\partial_x)D_{kl}(x,y)\right] = T_{il}(\partial_y)\left[A(\partial_x)D(x,y)\right]_{jl} = 0,$$

it follows that the columns $D^{(i)}$ and $P^{(i)}$ of D and P, respectively, are solutions of (2.11) at all points $x \neq y$.

We are now ready to state the analogue of Green's representation theorem, which involves the so-called *Somigliana formulae*.

2.10. Theorem. (i) *If $u \in C^2(S^+) \cap C^1(\bar{S}^+)$ satisfies $(Au)(x) = 0$, $x \in S^+$, then*

$$\int_{\partial S}\left[D(x,y)(Tu)(y) - P(x,y)u(y)\right]ds(y) = \begin{cases} u(x) & x \in S^+, \\ \tfrac{1}{2}u(x) & x \in \partial S, \\ 0 & x \in S^-. \end{cases} \qquad (2.42)$$

(ii) *If $u \in C^2(S^-) \cap C^1(\bar{S}^-) \cap \mathcal{A}$ satisfies $(Au)(x) = 0$, $x \in S^-$, then*

$$-\int_{\partial S}\left[D(x,y)(Tu)(y) - P(x,y)u(y)\right]ds(y) = \begin{cases} 0 & x \in S^+, \\ \tfrac{1}{2}u(x) & x \in \partial S, \\ u(x) & x \in S^-. \end{cases} \qquad (2.43)$$

Proof. (i) First, let $x \in S^+$, and let $\sigma_{x,\varepsilon}$ be a disk with the centre at x and radius ε sufficiently small so that $\bar{\sigma}_{x,\varepsilon}$ lies strictly inside S^+. By (2.33) applied in $\bar{S}^+ \setminus \sigma_{x,\varepsilon}$ with u and v replaced by the columns $D^{(\alpha)}$ and u, respectively,

$$\left(\int_{\partial S} + \int_{\partial \sigma_{x,\varepsilon}}\right) \left[(D^{(\alpha)})^{\mathrm{T}}(y,x)(Tu)(y) - u^{\mathrm{T}}(y)T(\partial_y)D^{(\alpha)}(y,x)\right] ds(y) = 0.$$

In view of (2.38) and (2.39), the integrand above can be written in terms of components as

$$D_{\beta\alpha}(y,x)(Tu)_\beta(y) - u_\beta(y)\bigl(T(\partial_y)D^{(\alpha)}(y,x)\bigr)_\beta$$
$$= D_{\alpha\beta}^{\mathrm{T}}(y,x)(Tu)_\beta(y) - \bigl(T_{\beta\gamma}(\partial_y)D_{\gamma\alpha}(y,x)\bigr)u_\beta(y)$$
$$= D_{\alpha\beta}(x,y)(Tu)_\beta(y) - P_{\alpha\beta}(x,y)u_\beta(y);$$

therefore,

$$\left(\int_{\partial S} + \int_{\partial \sigma_{x,\varepsilon}}\right) \left[D(x,y)(Tu)(y) - P(x,y)u(y)\right] ds(y) = 0. \tag{2.44}$$

By (2.40) and the boundedness of the derivatives of u in $\bar{S}^+$,

$$\int_{\partial \sigma_{x,\varepsilon}} D(x,y)(Tu)(y)\, ds(y) = O(\varepsilon \ln \varepsilon) \to 0 \quad \text{as } \varepsilon \to 0.$$

Next, we write

$$\int_{\partial \sigma_{x,\varepsilon}} P(x,y)u(y)\, ds(y) = \int_{\partial \sigma_{x,\varepsilon}} P(x,y)\bigl[u(y) - u(x)\bigr] ds(y) + \left(\int_{\partial \sigma_{x,\varepsilon}} P(x,y)\, ds(y)\right) u(x).$$

Since in this case the normal on $\partial \sigma_{x,\varepsilon}$ is directed into $\sigma_{x,\varepsilon}$, we use (2.41) and polar coordinates ε, θ with the pole at x to see that, for $y \in \partial \sigma_{x,\varepsilon}$, the matrix P is of the form

$$(2\pi)^{-1}\varepsilon^{-1}E_2 + \varepsilon^{-1}\mathcal{G}(\theta),$$

where $\int\limits_0^{2\pi} \mathcal{G}(\theta)\, d\theta = 0$. Also, by the continuity of u in $\bar{S}^+$, we have $u(y) - u(x) = O(\varepsilon)$, $y \in \partial \sigma_{x,\varepsilon}$. Consequently,

$$\int_{\partial \sigma_{x,\varepsilon}} P(x,y)u(y)\, ds(y) = \int_0^{2\pi} O(\varepsilon^{-1}) \cdot O(\varepsilon^{-1})\varepsilon\, d\theta + \left(E_2 + \int_0^{2\pi} \mathcal{G}(\theta)\, d\theta\right) u(x),$$

which tends to $u(x)$ as $\varepsilon \to 0$. The first formula (2.42) now follows from (2.44).

If $x \in \partial S$, then only the approximation of half of the disk $\sigma_{x,\varepsilon}$ is needed to isolate the point. When $x \in S^-$, $D(x,y)$ is analytic at any $y \in S^+$ and the argument involving $\sigma_{x,\varepsilon}$ is not necessary.

(ii) Let $x \in S^-$, and let K_R be a disk with the centre at x and radius R sufficiently large so that $\bar{S}^+$ lies strictly inside K_R. Applying (2.42) in $\bar{S}^- \cap \bar{K}_R$, we find that

$$\left(-\int_{\partial S} + \int_{\partial K_R}\right) \left[D(x,y)(Tu)(y) - P(x,y)u(y)\right] ds(y) = u(x),$$

since here the outward normal on ∂S is directed into S^+. If we switch to polar coordinates with the pole at x, then, by (1.12), (2.40), (2.41), (2.22) and (2.24), for R large the quantity between square brackets above is $O(R^{-2} \ln R)$, so the integral over ∂K_R vanishes as $R \to \infty$ and we obtain $(2.43)_3$. The cases $x \in \partial S$ and $x \in S^+$ are treated as in (i).

2.4. Uniqueness theorems

These assertions form an important step in the solution of the boundary value problems stated in §2.3.

2.11. Theorem. (i) (D^+), (D^-), (N^-), (R^+) *and* (R^-) *have at most one solution.*
(ii) *Any two solutions of* (N^+) *differ by a rigid displacement* Fk.

Proof. The difference u of any two solutions of (D^+) satisfies the same problem with homogeneous boundary conditions, so, by (2.31),

$$\int_{S^+} \mathcal{E}(u,u)\, da = 0.$$

Since $u \in C^1(\bar{S}^+)$ and $\mathcal{E}$ is a positive quadratic form (Theorem 2.2), this yields

$$\mathcal{E}(u,u) = 0, \qquad (2.45)$$

which, in turn (again by Theorem 2.2), implies that u is a rigid displacement. The homogeneous boundary condition for u now dictates that $u = 0$.

The reasoning is similar for (D^-), where use is made of (2.32).

In (N^+), u satisfies (2.45); consequently, it is an arbitrary rigid displacement. In (N^-), since $u \in \mathcal{A}$, the rigid displacement must be zero.

For (R^+) we write

$$(Tu)(x) = -(\sigma u)(x), \quad x \in \partial S,$$

and (2.31) leads to

$$\int_{S^+} \mathcal{E}(u,u)\,da + \int_{\partial S} u^{\mathrm{T}}(\sigma u)\,ds = 0.$$

Since σ is positive definite, we deduce that u is a rigid displacement in S^+ which vanishes on ∂S, so $u = 0$.

The argument is analogous in (R^-), with

$$(Tu)(x) = (\sigma u)(x), \quad x \in \partial S,$$

replaced in (2.32).

2.12. Corollary. (i) *If $u^{\mathcal{A}} + Fk$, where $k \in \mathcal{M}_{3\times 1}$ is constant, is a solution of the homogeneous problem* (D^-) *or* (R^-), *then $k = 0$ and $u^{\mathcal{A}} = 0$.*

(ii) *If $u \in \mathcal{A}$ is a solution of (D^-) with $u|_{\partial S} = Fk$, where $k \in \mathcal{M}_{3\times 1}$ is constant, then $k = 0$ and $u = 0$.*

Proof. (i) By Theorem 2.11(i), $u^{\mathcal{A}} + Fk = 0$ is the only solution in $\mathcal{A}^*$ of (D^+) or (R^+) with homogeneous boundary data. Then $Fk = -u^{\mathcal{A}} \in \mathcal{A}$, so $Fk = 0$. Since the columns $F^{(i)}$ of F are linearly independent, we deduce that $k = 0$, which implies that $u^{\mathcal{A}} = 0$.

(ii) We apply (i) to $u - Fk$.

2.5. The elastic potentials

The elastic potentials are constructed by means of the matrices $D(x,y)$ and $P(x,y)$ of singular solutions introduced in §2.3.

2.13. Definition. The single layer potential and double layer potential are defined by

$$(V\varphi)(x) = \int_{\partial S} D(x,y)\varphi(y)\,ds(y),$$

$$(W\psi)(x) = \int_{\partial S} P(x,y)\psi(y)\,ds(y),$$

where $\varphi, \psi \in \mathcal{M}_{2\times 1}$ are density functions with certain smoothness properties.

2.14. Theorem. *If $\varphi, \psi \in C(\partial S)$, then*
(i) $V\varphi \in \mathcal{A}$ *if and only if $p\varphi = 0$;*
(ii) $W\psi \in \mathcal{A}$.

Proof. Computing additional terms in (1.12), we find that, as $|x| \to \infty$,

$$\ln|x - y| = \ln|x| - (x_1 y_1 + x_2 y_2)|x|^{-2} + O(|x|^{-2}), \qquad (2.46)$$
$$|x - y|^{-2} = |x|^{-2} + 2(x_1 y_1 + x_2 y_2)|x|^{-4} + O(|x|^{-3}).$$

Using (2.46) and (2.26) in (2.40) and (2.41), we arrive at the expansions

$$(V\varphi)(r, \theta) = M^\infty(r, \theta)(p\varphi) + (V\varphi)^{\mathcal{A}}(r, \theta), \qquad (2.47)$$
$$(W\psi)(r, \theta) = (W\psi)^{\mathcal{A}}(r, \theta),$$

where

$$4\pi\mu(\gamma + 1)M^\infty(r, \theta)$$
$$= \begin{pmatrix} -2\gamma(\ln r + 1) + \cos 2\theta & \sin 2\theta & r^{-1}(\gamma + 1)\sin\theta \\ \sin 2\theta & -2\gamma(\ln r + 1) - \cos 2\theta & -r^{-1}(\gamma + 1)\cos\theta \end{pmatrix} \qquad (2.48)$$

and $(V\varphi)^{\mathcal{A}}$ and $(W\psi)^{\mathcal{A}}$ fit the pattern (2.22) with the coefficients

$$m_0 = \left[8\pi\mu(\lambda + 2\mu)\right]^{-1}(\lambda + \mu)\int_{\partial S}(y_1\varphi_2 + y_2\varphi_1)\,ds,$$

$$m_1 = \left[8\pi\mu(\lambda + 2\mu)\right]^{-1}\left[4\mu\int_{\partial S}y_1\varphi_1\,ds + (\lambda + \mu)\int_{\partial S}(y_1\varphi_1 - y_2\varphi_2)\,ds\right],$$

$$m_3 = \left[8\pi\mu(\lambda + 2\mu)\right]^{-1}\left[4\mu\int_{\partial S}y_2\varphi_2\,ds - (\lambda + \mu)\int_{\partial S}(y_1\varphi_1 - y_2\varphi_2)\,ds\right],$$

$$m_2 = m_4 = \left[8\pi\mu(\lambda + 2\mu)\right]^{-1}(\lambda + \mu)\int_{\partial S}(y_1\varphi_1 - y_2\varphi_2)\,ds$$

for the former and

$$m_0 = \left[4\pi(\lambda + 2\mu)\right]^{-1}(\lambda + \mu)\int_{\partial S}(\nu_2\psi_1 + \nu_1\psi_2)\,ds,$$

$$m_1 = \left[4\pi(\lambda + 2\mu)\right]^{-1}\left[4(\lambda + \mu)\int_{\partial S}\nu_1\psi_1\,ds + (\mu - \lambda)\int_{\partial S}(\nu_1\psi_1 - \nu_2\psi_2)\,ds\right],$$

$$m_3 = \left[4\pi(\lambda + 2\mu)\right]^{-1}\left[4(\lambda + \mu)\int_{\partial S}\nu_2\psi_2\,ds - (\mu - \lambda)\int_{\partial S}(\nu_1\psi_1 - \nu_2\psi_2)\,ds\right],$$

$$m_2 = m_4 = \left[4\pi(\lambda + 2\mu)\right]^{-1}(\lambda + \mu)\int_{\partial S}(\nu_1\psi_1 - \nu_2\psi_2)\,ds$$

for the latter. The assertion now follows immediately from (2.47).

THE ELASTIC POTENTIALS

2.15. Lemma. $(AM^\infty)(x) = 0$, $x \in S^-$.

Proof. Changing back from polar to Cartesian coordinates and taking (2.23) into account, from (2.48) we readily obtain

$$M_{11}^\infty(x) = [8\pi\mu(\lambda + 2\mu)]^{-1}\big[-2(\lambda + 3\mu)(\ln|x| + 1) + (\lambda + \mu)(x_1^2 - x_2^2)|x|^{-2}\big],$$
$$M_{22}^\infty(x) = [8\pi\mu(\lambda + 2\mu)]^{-1}\big[-2(\lambda + 3\mu)(\ln|x| + 1) - (\lambda + \mu)(x_1^2 - x_2^2)|x|^{-2}\big],$$
$$M_{12}^\infty(x) = M_{21}^\infty(x) = [4\pi\mu(\lambda + 2\mu)]^{-1}(\lambda + \mu)x_1 x_2 |x|^{-2},$$
$$M_{13}^\infty(x) = (4\pi\mu)^{-1}x_2|x|^{-2}, \quad M_{23}^\infty(x) = (4\pi\mu)^{-1}x_1|x|^{-2}.$$

The desired equality can now be verified directly.

2.16. Theorem. (i) If $\varphi, \psi \in C(\partial S)$, then $(V\varphi)(x)$ and $(W\psi)(x)$ are analytic at all $x \in S^+ \cup S^-$ and

$$A(V\varphi)(x) = A(W\psi)(x) = 0, \quad x \in S^+ \cup S^-.$$

(ii) If $\varphi \in C^{0,\alpha}(\partial S)$, then the direct values $V_0\varphi$ and $W_0\psi$ of $V\varphi$ and $W\psi$ on ∂S exist (the latter in the sense of principal value). Also, the operators $\mathcal{V}^\pm$ defined by

$$\mathcal{V}^+\varphi = (V\varphi)|_{\bar{S}^+}, \quad \mathcal{V}^-\varphi = (V\varphi)|_{\bar{S}^-}$$

map $C^{0,\alpha}(\partial S)$ to $C^{1,\alpha}(\bar{S}^\pm)$, $\alpha \in (0,1)$, respectively, and

$$T(\mathcal{V}^+\varphi) = (W_0^* + \tfrac{1}{2}I)\varphi, \quad T(\mathcal{V}^-\varphi) = (W_0^* - \tfrac{1}{2}I)\varphi, \quad \varphi \in C^{0,\alpha}(\partial S), \qquad (2.49)$$

where I is the identity operator and W_0^* is the adjoint of the direct value operator W_0, defined (in the sense of principal value) by

$$(W_0^*\varphi)(x) = \int_{\partial S} \big(T(\partial_x)D(x,y)\big)\varphi(y)\,ds(y), \quad x \in \partial S.$$

(iii) The operators $\mathcal{W}^\pm$ defined by

$$\mathcal{W}^+\psi = \begin{cases} (W\psi)|_{S^+} & \text{in } S^+, \\ (W_0 - \tfrac{1}{2}I)\psi & \text{on } \partial S, \end{cases} \quad \mathcal{W}^-\psi = \begin{cases} (W\psi)|_{S^-} & \text{in } S^-, \\ (W_0 + \tfrac{1}{2}I)\psi & \text{on } \partial S \end{cases} \qquad (2.50)$$

map $C^{0,\alpha}(\partial S)$ to $C^{0,\alpha}(\bar{S}^\pm)$ and $C^{1,\alpha}(\partial S)$ to $C^{1,\alpha}(\bar{S}^\pm)$, $\alpha \in (0,1)$, respectively, and

$$T(\mathcal{W}^+\psi) = T(\mathcal{W}^-\psi), \quad \psi \in C^{1,\alpha}(\partial S). \qquad (2.51)$$

(iv) The operator W_0 maps $C^{0,\alpha}(\partial S)$ to $C^{1,\alpha}(\partial S)$, $\alpha \in (0,1)$.

The proof of this assertion follows from Theorems 1.3 and 1.4 since, by (2.40) and (2.41), the potentials can be written as

$$V\varphi = -\big[4\pi\mu(\gamma+1)\big]^{-1}\big[2\gamma(v\varphi) + (2\gamma+1)(\bar{p}\varphi) - 2v_{\alpha\beta}^{b}(E_{\alpha\beta}\varphi)\big],$$
$$W\psi = -(2\pi)^{-1}\big[(\gamma-1)(\gamma+1)^{-1}v^{f}(\varepsilon_{\alpha\beta}E_{\alpha\beta}\psi) + w\psi - 2(\gamma+1)^{-1}v_{\alpha\beta}^{e}(\varepsilon_{\alpha\delta}E_{\delta\beta}\psi)\big],$$

where

$$\bar{p}\varphi = \big((p\varphi)_1, (p\varphi)_2\big)^{\mathrm{T}} = \bigg(\int_{\partial S}\varphi_1\,ds, \int_{\partial S}\varphi_2\,ds\bigg)^{\mathrm{T}}.$$

2.17. Remarks. (i) The derivatives on ∂S of the functions defined in $\bar{S}^+$ or $\bar{S}^-$ in Theorem 2.16 are one-sided.

(ii) $A(\mathcal{V}^+\varphi)(x) = 0$, $x \in S^+$, and $A(\mathcal{V}^-\varphi)(x) = 0$, $x \in S^-$.

(iii) If $\mathcal{V}^+\varphi = \mathcal{V}^-\varphi = 0$, then, by (2.49), $\varphi = 0$.

(iv) The Somigliana relations (2.40) and (2.41) can be rewritten in the form

$$\mathcal{V}^+(Tu) - \mathcal{W}^+(u|_{\partial S}) = u, \qquad \mathcal{V}^-(Tu) - \mathcal{W}^-(u|_{\partial S}) = 0, \qquad (2.52)$$
$$-\mathcal{V}^-(Tu) + \mathcal{W}^-(u|_{\partial S}) = u, \qquad -\mathcal{V}^+(Tu) + \mathcal{W}^+(u|_{\partial S}) = 0. \qquad (2.53)$$

(v) From (2.52) with $u = F^{(i)}$ we derive the equalities

$$\mathcal{W}^+F = -F, \quad \mathcal{W}_0 F = -\tfrac{1}{2}F, \quad \mathcal{W}^-F = 0.$$

They do not contradict (2.53) since the latter are valid for $u \in \mathcal{A}$, and $F \notin \mathcal{A}$.

(vi) Formula (2.51) makes it possible to define a boundary operator

$$N_0 : C^{1,\alpha}(\partial S) \to C^{0,\alpha}(\partial S)$$

by setting

$$N_0\psi = T(\mathcal{W}^+\psi) = T(\mathcal{W}^-\psi), \quad \psi \in C^{1,\alpha}(\partial S). \qquad (2.54)$$

(vii) In view of Theorem 2.16, we adopt the notation

$$(\mathcal{V}^\pm\varphi)|_{\partial S} = \mathcal{V}_0^\pm\varphi = V_0\varphi, \quad (\mathcal{W}^\pm\psi)|_{\partial S} = \mathcal{W}_0^\pm\psi = (W_0 \mp \tfrac{1}{2}I)\psi. \qquad (2.55)$$

For simplicity, we also write certain equations in the style $A\mathcal{U}^+ = 0$ as an abbreviation for $(A\mathcal{U}^+)(x) = 0$, $x \in S^+$, when $\mathcal{U}^+$ is a function defined in $\bar{S}^+$. The same convention is used in the case of $\bar{S}^-$, ∂S and $\mathbb{R}^2$.

Unless stated otherwise, in the rest of this chapter we assume that the boundary integral operators V_0, W_0 and W_0^* and the combinations $W_0 \pm \tfrac{1}{2}I$ and $W_0^* \pm \tfrac{1}{2}I$ are defined on $C^{0,\alpha}(\partial S)$, while N_0 is defined on $C^{1,\alpha}(\partial S)$, $\alpha \in (0,1)$.

2.6. Properties of the boundary operators

We start with the algebra of the boundary operators. If the proof of the corresponding assertion for the Laplace equation (Theorem 1.36) seemed a little too convoluted, then here is an alternative.

2.18. Theorem. V_0, W_0, W_0^* and N_0 satisfy the composition formulae

$$W_0 V_0 = V_0 W_0^*, \quad N_0 V_0 = W_0^{*2} - \tfrac{1}{4}I \quad \text{on } C^{0,\alpha}(\partial S), \tag{2.56}$$

$$N_0 W_0 = W_0^* N_0, \quad V_0 N_0 = W_0^2 - \tfrac{1}{4}I \quad \text{on } C^{1,\alpha}(\partial S). \tag{2.57}$$

Proof. Consider the function

$$u = \mathcal{V}^+ \psi - \mathcal{W}^+ \varphi, \tag{2.58}$$

where φ is arbitrary in $C^{1,\alpha}(\partial S)$ and ψ is arbitrary in $C^{0,\alpha}(\partial S)$. Clearly, u satisfies $(Au)(x) = 0$, $x \in S^+$. Setting

$$u|_{\partial S} = \alpha, \quad Tu = \beta,$$

restricting (2.58) to ∂S and using (2.55), we can write

$$\alpha = V_0 \psi - (W_0 - \tfrac{1}{2}I)\varphi. \tag{2.59}$$

Now taking the normal derivative (on ∂S) in (2.58) and using (2.49) and (2.54), we arrive at

$$\beta = (W_0^* + \tfrac{1}{2}I)\psi - N_0 \varphi. \tag{2.60}$$

On the other hand, applying the Somigliana representation formula $(2.52)_1$ to u and restricting it to ∂S, we obtain

$$\alpha = V_0 \beta - (W_0 - \tfrac{1}{2}I)\alpha. \tag{2.61}$$

Finally, as above, taking the normal derivative in the restriction of $(2.52)_1$ to S^+, we find that

$$\beta = (W_0^* + \tfrac{1}{2}I)\beta - N_0 \alpha. \tag{2.62}$$

If we define the matrices

$$U = \begin{pmatrix} \alpha \\ \beta \end{pmatrix}, \quad \Theta = \begin{pmatrix} \varphi \\ \psi \end{pmatrix}, \quad Q = \begin{pmatrix} -(W_0 - \tfrac{1}{2}I) & V_0 \\ -N_0 & W_0^* + \tfrac{1}{2}I \end{pmatrix},$$

72   PLANE STRAIN

then (2.59)–(2.62) can be written as

$$U = Q\Theta, \quad U = QU.$$

Substituting U from the first of these equalities into the second one leads to the equation $(Q^2 - Q)\Theta = 0$. Given the arbitrariness of Θ in $C^{1,\alpha}(\partial S) \times C^{0,\alpha}(\partial S)$, this implies that $Q^2 = Q$, which is equivalent to (2.56) and (2.57).

2.19. Theorem. *The operators $W_0 \pm \tfrac{1}{2}I$ and $W_0^* \pm \tfrac{1}{2}I$ are of index zero.*

Proof. If we set $z = x_1 + ix_2$ and $\zeta = y_1 + iy_2$, then

$$\log(\zeta - z) = \ln|\zeta - z| + i\theta = \ln|x - y| + i\theta, \quad \theta = \arg(\zeta - z).$$

Differentiating this equality along ∂S with respect to the arc element at y and using the Cauchy-Riemann equation

$$\partial_{s(y)}\theta(x, y) = \partial_{\nu(y)} \ln|x - y|,$$

we find that

$$|\zeta - z|^{-1} d\zeta = (\partial_{s(y)} \ln|x - y|)\,ds(y) + i(\partial_{\nu(y)} \ln|x - y|)\,ds(y).$$

We now replace $(\partial_{s(y)} \ln|x - y|)\,ds(y)$ from this in (2.41) and write

$$(W_0 - \tfrac{1}{2}I)\varphi = K^s\varphi + K^w\varphi - \tfrac{1}{2}\varphi,$$

where

$$(K^s\varphi)(z) = -(2\pi)^{-1}(\gamma - 1)(\gamma + 1)^{-1}\varepsilon_{\alpha\beta}E_{\alpha\beta} \int_{\partial S} (\zeta - z)^{-1}\varphi(\zeta)\,d\zeta, \quad z \in \partial S,$$

and K^w is a weakly singular operator (see Remark 1.8). It is easy to check that the properties required in Definition 1.16 are satisfied with

$$\hat{k}^s(z, z) = -(2\pi)^{-1}(\gamma - 1)(\gamma + 1)^{-1}\varepsilon_{\alpha\beta}E_{\alpha\beta}, \quad z \in \partial S,$$

so K^s is α-regular singular, $\alpha \in (0, 1)$. Also, by Remark 1.18(ii), K^w is α-regular singular and $\hat{k}^w(z, z) = 0$. Consequently, $W_0 - \tfrac{1}{2}I$ is α-regular singular and, with γ given by (2.23),

$$\hat{k}(z, z) = \hat{k}^s(z, z) + \hat{k}^w(z, z) = -(2\pi)^{-1}\mu(\lambda + 2\mu)^{-1}\varepsilon_{\alpha\beta}E_{\alpha\beta}, \quad z \in \partial S;$$

therefore,

$$\det\left(-\tfrac{1}{2}E_2 \pm \pi i \hat{k}(z,z)\right) = \tfrac{1}{4}(\lambda+\mu)(\lambda+3\mu)(\lambda+2\mu)^{-2} \neq 0 \quad \text{for all } z \in \partial S.$$

This means that the index of $W_0 - \tfrac{1}{2}I$, given by (1.4), is $\rho = 0$. Similarly, the index of $W_0 + \tfrac{1}{2}I$ is also zero, and, in view of Remark 1.18(i), so are those of $W_0^* \pm \tfrac{1}{2}I$.

2.20. Corollary. *The Fredholm Alternative holds for the pairs of equations*

$$(W_0 - \tfrac{1}{2}I)\varphi = f, \quad (W_0^* - \tfrac{1}{2}I)\psi = g, \quad f, g \in C^{0,\alpha}(\partial S),$$
$$(W_0 + \tfrac{1}{2}I)\varphi = f, \quad (W_0^* + \tfrac{1}{2}I)\psi = g, \quad f, g \in C^{0,\alpha}(\partial S),$$

in the dual system $\bigl(C^{0,\alpha}(\partial S), C^{0,\alpha}(\partial S)\bigr)$, $\alpha \in (0,1)$, with the bilinear form

$$(\varphi, \psi) = \int_{\partial S} \varphi^{\mathrm{T}} \psi \, ds.$$

The proof of this assertion follows immediately from Theorems 1.19 and 2.19.

2.21. Theorem. *If there is $\varphi \in C^{0,\alpha}(\partial S)$, $\varphi \neq 0$, such that*

$$(W_0^* + \tfrac{1}{2}I)\varphi = 0,$$

then

$$\mathcal{V}^+\varphi = Fk$$

for some constant $k \in \mathcal{M}_{3\times 1}$, and

$$p\varphi \neq 0.$$

Proof. By (2.49), $(W_0^* + \tfrac{1}{2}I)\varphi = 0$ is equivalent to $T(\mathcal{V}^+\varphi) = 0$, which implies that $\mathcal{V}^+\varphi$ is a solution of the homogeneous problem (N$^+$). By Theorem 2.11(ii), $\mathcal{V}^+\varphi$ is a rigid displacement Fk.

If $p\varphi = 0$, then the equality

$$\mathcal{V}_0^-\varphi = V_0\varphi = \mathcal{V}_0^+\varphi = Fk$$

means that $\mathcal{V}^- \in \mathcal{A}$ is a solution of (D$^-$) with boundary data Fk, so, by Corollary 2.12(ii), $Fk = \mathcal{V}^-\varphi = 0$. Consequently, we also have $\mathcal{V}^+\varphi = Fk = 0$, and we use Remark 2.17(iii) to deduce that $\varphi = 0$, which contradicts the assumption in the theorem. Hence, $p\varphi \neq 0$.

2.22. Theorem. (i) *The null spaces of $W_0 - \tfrac{1}{2}I$ and $W_0^* - \tfrac{1}{2}I$ consist of the zero vector alone.*

(ii) *The null spaces of $W_0 + \tfrac{1}{2}I$ and $W_0^* + \tfrac{1}{2}I$ are three-dimensional and are spanned, respectively, by $\{F^{(i)}\}$ and the (linearly independent) columns $\{\Phi^{(i)}\}$ of a (2×3)-matrix $\Phi \in C^{0,\alpha}(\partial S)$.*

Proof. (i) If $\psi_0 \in C^{0,\alpha}(\partial S)$ is in the null space of $W_0^* - \tfrac{1}{2}I$, then, by (2.39) and Remark 2.17(v),

$$\begin{aligned}
0 &= p(W_0^* \psi_0) - \tfrac{1}{2}(p\psi_0) \\
&= \int_{\partial S} F^{\mathrm{T}}(x) \left(\int_{\partial S} P(x,y) \psi_0(y)\, ds(y) \right) ds(x) - \tfrac{1}{2}(p\psi_0) \\
&= \int_{\partial S} \left(\int_{\partial S} F^{\mathrm{T}}(x) \left[T(\partial_y) D(y,x) \right]^{\mathrm{T}} ds(x) \right) \psi_0(y)\, ds(y) - \tfrac{1}{2}(p\psi_0) \\
&= \int_{\partial S} \left(\int_{\partial S} \left(T(\partial_y) D(y,x) \right) F(x)\, ds(x) \right)^{\mathrm{T}} \psi_0(y)\, ds(y) - \tfrac{1}{2}(p\psi_0) \\
&= \int_{\partial S} (W_0 F)^{\mathrm{T}} \psi_0\, ds - \tfrac{1}{2}(p\psi_0) = -\tfrac{1}{2} \int_{\partial S} F^{\mathrm{T}} \psi_0\, ds - \tfrac{1}{2}(p\psi_0) = -p\psi_0.
\end{aligned}$$

This means that, by Theorem 2.14(i), $\mathcal{V}^- \psi_0 \in \mathcal{A}$. Since

$$T(\mathcal{V}^- \psi_0) = (W_0^* - \tfrac{1}{2}I)\psi_0 = 0, \quad A(\mathcal{V}^- \psi_0) = 0,$$

it follows that $\mathcal{V}^- \psi_0$ is a solution of the homogeneous problem (N^-), so $\mathcal{V}^- \psi_0 = 0$. Then

$$\mathcal{V}_0^+ \psi_0 = V_0 \psi_0 = \mathcal{V}_0^- \psi_0,$$

which means that $\mathcal{V}^+ \varphi_0 = 0$ as the unique solution of the homogeneous problem (D^+). By Remark 2.33(iii), $\psi_0 = 0$. We now use Corollary 2.20 and the Fredholm Alternative to conclude that the null space of $W_0 - \tfrac{1}{2}I$ also consists of the zero vector alone.

(ii) Since, by Theorem 2.16(iv), any $\varphi_0 \in C^{0,\alpha}(\partial S)$ satisfying $(W_0 + \tfrac{1}{2}I)\varphi_0 = 0$ belongs to $C^{1,\alpha}(\partial S)$, we prove the assertion for $W_0 + \tfrac{1}{2}I$ in $C^{1,\alpha}(\partial S)$.

We know that any rigid displacement Fk satisfies

$$A(Fk) = 0, \quad T(Fk) = 0,$$

so from the Somigliana formula $(2.52)_1$ restricted to ∂S we find that

$$-(W_0 - \tfrac{1}{2}I)(Fk) = Fk,$$

or

$$(W_0 + \tfrac{1}{2}I)(Fk) = 0$$

for any constant $k \in \mathcal{M}_{3\times 1}$. Consequently, the columns $F^{(i)}$ of F belong to the null space of $W_0 + \tfrac{1}{2}I$. Since the $F^{(i)}$ are linearly independent, it follows that the dimension of this space is at least 3.

Let $f^{(0)} \in C^{1,\alpha}(\partial S)$ be arbitrary in the null space of $W_0 + \tfrac{1}{2}I$. Then for any constant $c \in \mathcal{M}_{3\times 1}$, the $C^{1,\alpha}$-function $f^{(0)} - Fc$ also belongs to the null space of $W_0 + \tfrac{1}{2}I$; in other words,

$$(W_0 + \tfrac{1}{2}I)(f^{(0)} - Fc) = 0.$$

Since this is equivalent to $\mathcal{W}_0^-(f^{(0)} - Fc) = 0$, and since $\mathcal{W}^-(f^{(0)} - Fc) \in \mathcal{A}$, it follows that $\mathcal{W}^-(f^{(0)} - Fc)$ is a solution of the homogeneous problem (D^-); therefore,

$$\mathcal{W}^-(f^{(0)} - Fc) = 0. \tag{2.63}$$

Then, by (2.51) and (2.63),

$$T\mathcal{W}^-(f^{(0)} - Fc) = 0 = T\mathcal{W}^+(f^{(0)} - Fc),$$

which implies that $\mathcal{W}^+(f^{(0)} - Fc)$ is a solution of the homogeneous problem (N^+). Hence, by Theorem 2.11(ii) and Remark 2.17(v),

$$\mathcal{W}^+(f^{(0)} - Fc) = \mathcal{W}^+ f^{(0)} - (\mathcal{W}^+ F)c = \mathcal{W}^+ f^{(0)} + Fc = Fk \tag{2.64}$$

for some constant $k \in \mathcal{M}_{3\times 1}$. Recalling that the origin is in S^+, we now choose c to be the (constant) vector of components

$$c_\alpha = -(\mathcal{W}^+ f^{(0)})_\alpha(0,0), \quad c_3 = -\tfrac{1}{2}\varepsilon_{\alpha\beta}(\mathcal{W}^+ f^{(0)})_{\alpha,\beta}(0,0). \tag{2.65}$$

(This choice is possible since, by Theorem 2.16(iii), $\mathcal{W}^+ f^{(0)} \in C^{1,\alpha}(\bar{S}^+)$.) Considering (2.64) at the origin, we find that

$$k_\alpha = (\mathcal{W}^+ f^{(0)})_\alpha(0,0) + c_\alpha = 0.$$

Since
$$\varepsilon_{\alpha\beta}(Fk)_{\alpha,\beta} = (Fk)_{1,2} - (Fk)_{2,1} = 2k_3,$$
equality (2.64) also yields
$$k_3 = \tfrac{1}{2}\left[\varepsilon_{\alpha\beta}(\mathcal{W}^+ f^{(0)})_{\alpha,\beta}(0,0) + 2c_3\right] = 0.$$
Consequently, (2.64) reduces to
$$\mathcal{W}^+(f^{(0)} - Fc) = 0. \tag{2.66}$$

Restricting (2.63) and (2.66) to ∂S and using (2.50), we conclude that $f^{(0)} - Fc = 0$; that is, $f^{(0)}$ is a linear combination of the $F^{(i)}$ with coefficients given by (2.65), which means that the null space of $W_0 + \tfrac{1}{2}I$ is three-dimensional and is spanned by $\{F^{(i)}\}$.

By the Fredholm Alternative, the null space of $W_0^* + \tfrac{1}{2}I$ is also three-dimensional and is spanned by the (linearly independent) columns $\Phi^{(i)}$ of a (2×3)-matrix $\Phi \in C^{0,\alpha}(\partial S)$.

2.23. Remark. Clearly, Φ is not unique. A little later we will choose Φ uniquely by asking it to satisfy a certain convenient calibration.

2.24. Theorem. (i) *The null spaces of $W_0^2 - \tfrac{1}{4}I$ and $W_0^{*2} - \tfrac{1}{4}I$ coincide with those of $W_0 + \tfrac{1}{2}I$ and $W_0^* + \tfrac{1}{2}I$ (that is, they are spanned by $\{F^{(i)}\}$ and $\{\Phi^{(i)}\}$, respectively).*

(ii) $N_0\psi = 0$ *if and only if* $\psi = Fk$, *where* $k \in \mathcal{M}_{3\times 1}$ *is constant.*

Proof. (i) In view of the mapping properties of W_0 and W_0^* in Theorem 2.16, the compositions
$$W_0^2 - \tfrac{1}{4}I = (W_0 - \tfrac{1}{2}I)(W_0 + \tfrac{1}{2}I), \quad W_0^{*2} - \tfrac{1}{4}I = (W_0^* - \tfrac{1}{2}I)(W_0^* + \tfrac{1}{2}I)$$
are legitimate. By Theorem 2.22(i), the equations
$$(W_0^2 - \tfrac{1}{4}I)\varphi = 0, \quad (W_0^{*2} - \tfrac{1}{4}I)\psi = 0$$
are equivalent, respectively, to
$$(W_0 + \tfrac{1}{2}I)\varphi = 0, \quad (W_0^* + \tfrac{1}{2}I)\psi = 0,$$
which proves the assertion.

(ii) If $\psi = Fk$, then, by Remark 2.17(v),

$$N_0\psi = T(\mathcal{W}^+(Fk)) = T(\mathcal{W}^+F)k = -(TF)k = 0.$$

If $N_0\psi = 0$, then, by $(2.57)_2$,

$$0 = V_0(N_0\psi) = (V_0N_0)\psi = (W_0^2 - \tfrac{1}{4}I)\psi,$$

and we apply (i) and Theorem 2.22(ii) to deduce that $\psi = Fk$ for some constant vector $k \in \mathcal{M}_{3\times 1}$.

As in Chapter 1, there is a more formal characterization of the boundary contour ∂S in terms of the matrix Φ.

2.25. Theorem. *For every simple closed C^2-curve ∂S and any $\alpha \in (0,1)$, there are a unique (2×3)-matrix $\Phi \in C^{0,\alpha}(\partial S)$ and a unique constant (3×3)-matrix $\mathcal{C}$ such that the columns $\Phi^{(i)}$ of Φ are linearly independent and*

$$V_0\Phi = F\mathcal{C}, \quad p\Phi = E_3. \tag{2.67}$$

Proof. From Theorems 2.22 and 2.24(i) we know that there is a (2×3)-matrix $\Phi \in C^{0,\alpha}(\partial S)$ with linearly independent columns $\Phi^{(i)}$ which satisfies

$$(W_0^{*2} - \tfrac{1}{4}I)\Phi = 0.$$

By $(2.56)_2$, this equation is equivalent to

$$(N_0V_0)\Phi = N_0(V_0\Phi) = 0,$$

which, according to Theorem 2.24(ii), implies that

$$V_0\Phi = FK \tag{2.68}$$

for some constant $K \in \mathcal{M}_{3\times 3}$.

Suppose that $\det(p\Phi) = 0$. Then the columns of $p\Phi$ are linearly dependent, so there is a constant non-zero $h \in \mathcal{M}_{3\times 1}$ such that

$$(p\Phi)h = 0. \tag{2.69}$$

By (2.47), (2.68) and (2.69),

$$A(\mathcal{V}^-(\Phi h) - FKh) = 0,$$
$$\mathcal{V}_0^-(\Phi h) - FKh = (V_0\Phi - FK)h = 0,$$
$$(\mathcal{V}^-(\Phi h) - FKh)(x) = (M^\infty(p\Phi)h + \mathcal{U}^\mathcal{A} - FKh)(x)$$
$$= (-FKh + \mathcal{U}^\mathcal{A})(x) \quad \text{as } |x| \to \infty.$$

This is the homogeneous problem (D$^-$), whose unique solution is

$$\mathcal{V}^-(\Phi h) - FKh = 0. \tag{2.70}$$

Next,

$$\mathcal{V}_0^+(\Phi h) - FKh = \mathcal{V}_0^-(\Phi h) - FKh = 0$$

means that

$$\mathcal{V}^+(\Phi h) - FKh = 0 \tag{2.71}$$

as the unique solution of the homogeneous problem (D$^+$). Since $TF = 0$, from (2.70), (2.71) and Remark 2.17(iii) it follows that $\Phi h = 0$, which contradicts the linear independence of the $\Phi^{(i)}$. Consequently, $\det(p\Phi) \neq 0$; therefore, the constant (3×3)-matrix $p\Phi$ is invertible. In this case, by (2.68),

$$V_0(\Phi(p\Phi)^{-1}) = (V_0\Phi)(p\Phi)^{-1} = FK(p\Phi)^{-1}$$

and

$$p(\Phi(p\Phi)^{-1}) = (p\Phi)(p\Phi)^{-1} = E_3,$$

so we can take $\Phi(p\Phi)^{-1}$ and $\mathcal{C} = K(p\Phi)^{-1}$ as a solution pair to our problem. We remark that the equality $\det(p\Phi) \neq 0$ implies that the columns of $\Phi(p\Phi)^{-1}$ are three linearly independent combinations of the $\Phi^{(i)}$, so they also form a basis for the null space of $W_0^* + \frac{1}{2}I$. Consequently, without loss of generality, we denote $\Phi(p\Phi)^{-1}$ by Φ and bear in mind that $p\Phi = E_3$.

The difference $\bar\Phi, \bar{\mathcal{C}}$ of any two solutions satisfies

$$A(\mathcal{V}^-\bar\Phi - F\bar{\mathcal{C}}) = 0,$$
$$\mathcal{V}_0^-\bar\Phi - F\bar{\mathcal{C}} = V_0\bar\Phi - F\bar{\mathcal{C}} = 0,$$
$$(\mathcal{V}^-\bar\Phi - F\bar{\mathcal{C}})(x) = (M^\infty(p\bar\Phi) + \mathcal{U}^\mathcal{A})(x) = \mathcal{U}^\mathcal{A}(x) \quad \text{as } |x| \to \infty.$$

By Corollary 2.12(i),

$$F\bar{\mathcal{C}} = 0, \quad \mathcal{V}^-\bar\Phi = 0.$$

Since the columns $F^{(i)}$ of F are linearly independent, we deduce that $\bar{C} = 0$. Also, the equality
$$\mathcal{V}_0^+ \bar{\Phi} = V_0 \bar{\Phi} = \mathcal{V}_0^- \bar{\Phi} = 0$$
implies that $\mathcal{V}^+ \bar{\Phi} = 0$ as the unique solution of the homogeneous problem (D$^+$); therefore, by Remark 2.17(iii), $\bar{\Phi} = 0$, which means that the solution pair Φ, $\mathcal{C}$ is unique.

2.26. Remarks. (i) The equality $p\Phi = E_3$ is equivalent to the biorthogonality of the sets $\{F^{(i)}\}$ and $\{\Phi^{(j)}\}$; that is,
$$(F^{(i)}, \Phi^{(j)}) = \int_{\partial S} (F^{(i)})^{\mathrm{T}} \Phi^{(j)} \, ds = \delta_{ij}, \quad i, j = 1, 2, 3.$$

(ii) Since (D$^+$) has at most one solution and $\mathcal{V}^+ \Phi$ and $F\mathcal{C}$ satisfy (D$^+$) and coincide on ∂S, it follows that $\mathcal{V}^+ \Phi = F\mathcal{C}$.

(iii) There seems to be no obvious connection between the Robin constant (or the logarithmic capacity of ∂S) and the matrix $\mathcal{C}$.

2.27. Example. We take ∂S to be the circle ∂K_R with the centre at the origin and radius R. By direct calculation we find that for all $x \in \partial K_R$

$$\int_{\partial K_R} \ln|x-y| \, ds(y) = 2\pi R \ln R,$$

$$\int_{\partial K_R} (x_\alpha - y_\alpha)^2 |x-y|^{-2} \, ds(y) = \pi R,$$

$$\int_{\partial K_R} (x_1 - y_1)(x_2 - y_2) |x-y|^{-2} \, ds(y) = 0,$$

$$\int_{\partial K_R} y_\alpha \ln|x-y| \, ds(y) = -\pi R x_\alpha,$$

$$\int_{\partial K_R} y_\alpha (x_\beta - y_\beta)^2 |x-y|^{-2} \, ds(y) = \tfrac{1}{2}\pi R x_\alpha, \quad \alpha \ne \beta,$$

$$\int_{\partial K_R} y_1 (x_1 - y_1)(x_2 - y_2) |x-y|^{-2} \, ds(y) = -\tfrac{1}{2}\pi R x_2,$$

$$\int_{\partial K_R} y_2 (x_1 - y_1)(x_2 - y_2) |x-y|^{-2} \, ds(y) = \tfrac{1}{2}\pi R x_1.$$

Then from (2.40) and (2.23) it follows that

$$\int_{\partial K_R} D_{11}(x,y)\,ds(y) = \int_{\partial K_R} D_{22}(x,y)\,ds(y) = \left[2\mu(\lambda+2\mu)\right]^{-1}(\lambda+3\mu)R(\ln R + 1),$$

$$\int_{\partial K_R} D_{12}(x,y)\,ds(y) = \int_{\partial K_R} D_{12}(x,y)\,ds(y) = 0,$$

$$\int_{\partial K_R} \left[y_2 D_{11}(x,y) - y_1 D_{12}(x,y)\right] ds(y) = (2\mu)^{-1} R x_2,$$

$$\int_{\partial K_R} \left[y_2 D_{21}(x,y) - y_1 D_{22}(x,y)\right] ds(y) = -(2\mu)^{-1} R x_1,$$

so

$$V_0 F = V_0 \begin{pmatrix} 1 & 0 & x_2 \\ 0 & 1 & -x_1 \end{pmatrix}$$

$$= -\left[2\mu(\lambda+2\mu)\right]^{-1} R \begin{pmatrix} (\lambda+3\mu)(\ln R + 1) & 0 & (\lambda+2\mu)x_2 \\ 0 & (\lambda+3\mu)(\ln R + 1) & -(\lambda+2\mu)x_1 \end{pmatrix}.$$

Also,

$$pF = \int_{\partial S} F^{\mathrm{T}} F\,ds = \int_{\partial S} \begin{pmatrix} 1 & 0 & x_2 \\ 0 & 1 & -x_1 \\ x_2 & -x_1 & x_1^2 + x_2^2 \end{pmatrix} ds = 2\pi R\,\mathrm{diag}\{1,1,R^2\};$$

therefore,

$$\left[4\pi\mu(\lambda+2\mu)R^2\right]V_0\bigl(F(pF)^{-1}\bigr) = \left[4\pi\mu(\lambda+2\mu)R^2\right](V_0 F)(pF)^{-1}$$

$$= \begin{pmatrix} (\lambda+3\mu)R^2(\ln R + 1) & 0 & (\lambda+2\mu)x_2 \\ 0 & (\lambda+3\mu)R^2(\ln R + 1) & -(\lambda+2\mu)x_1 \end{pmatrix} = F\mathcal{C},$$

where

$$\mathcal{C} = \left[4\pi\mu(\lambda+2\mu)R^2\right]^{-1} \mathrm{diag}\{(\lambda+3\mu)R^2(\ln R + 1), (\lambda+3\mu)R^2(\ln R + 1), (\lambda+2\mu)\}.$$

Since $p\bigl(F(pF)^{-1}\bigr) = E_3$, we have $\Phi = F(pF)^{-1}$. As mentioned in Remark 2.26(iii), no meaningful connection is apparent between $\mathcal{C}$ and the logarithmic capacity of ∂K_R, which is R (see Remark 1.41(ii)).

2.28. Corollary. *If $V_0 \varphi = Fc$ with a constant $c \in \mathcal{M}_{3\times 1}$, then*

$$\varphi = \Phi k, \quad c = \mathcal{C}k$$

for some constant $k \in \mathcal{M}_{3\times 1}$.

PROPERTIES OF THE BOUNDARY OPERATORS

Proof. Using the composition relation $(2.56)_2$, we find that, by Theorem 2.24(ii),

$$(W_0^{*2} - \tfrac{1}{4}I)\varphi = (N_0 V_0)\varphi = N_0(V_0\varphi) = N_0(Fc) = 0.$$

Since the $\Phi^{(i)}$ form a basis for the null space of $W_0^{*2} - \tfrac{1}{4}I$, this means that φ is a linear combination of the $\Phi^{(i)}$; in other words, $\varphi = \Phi k$ for some constant $k \in \mathcal{M}_{3\times 1}$. Then, by Theorem 2.25,

$$Fc = V_0\varphi = V_0(\Phi k) = (V_0\Phi)k = (FC)k = F(Ck),$$

or

$$F(c - Ck) = 0.$$

In view of the linear independence of the $F^{(i)}$, we conclude that $c = Ck$.

2.29. Theorem. *The equation $V_0\varphi = 0$ has non-zero solutions if and only if ∂S is such that $\det \mathcal{C} = 0$.*

Proof. Suppose that $\det \mathcal{C} = 0$. Then the columns of $\mathcal{C}$ are linearly dependent, so there is a constant $k \in \mathcal{M}_{3\times 1}$, $k \neq 0$, such that $\mathcal{C}k = 0$. By Theorem 2.25,

$$V_0(\Phi k) = (V_0\Phi)k = (F\mathcal{C})k = F(\mathcal{C}k) = 0.$$

Since the columns of Φ are linearly independent and $k \neq 0$, it follows that $\Phi k \neq 0$, which means that the equation $V_0\varphi = 0$ has non-zero solutions.

Suppose now that $\det \mathcal{C} \neq 0$, and that φ satisfies $V_0\varphi = 0$. By Theorem 2.25 and expansion (2.47),

$$A\bigl(\mathcal{V}^-\varphi + (F\mathcal{C} - \mathcal{V}^-\Phi)(p\varphi)\bigr) = 0,$$
$$V_0^-\varphi + (F\mathcal{C} - V_0^-\Phi)(p\varphi) = V_0\varphi + (F\mathcal{C} - V_0\Phi)(p\varphi) = 0,$$
$$\bigl(\mathcal{V}^-\varphi + (F\mathcal{C} - \mathcal{V}^-\Phi)(p\varphi)\bigr)(x)$$
$$= \bigl(M^\infty(p\varphi) + (V\varphi)^{\mathcal{A}} + (F\mathcal{C})(p\varphi) - M^\infty(p\Phi)(p\varphi) - (V\Phi)^{\mathcal{A}}(p\varphi)\bigr)(x)$$
$$= \bigl(F(\mathcal{C}(p\varphi)) + \mathcal{U}^{\mathcal{A}}\bigr)(x) \quad \text{as } |x| \to \infty.$$

By Corollary 2.12(i), this homogeneous problem (D$^-$) yields

$$F\bigl(\mathcal{C}(p\varphi)\bigr) = 0, \quad \mathcal{V}^-\varphi + (F\mathcal{C} - \mathcal{V}^-\Phi)(p\varphi) = 0.$$

Since the columns of F are linearly independent, it follows that $\mathcal{C}(p\Phi) = 0$, and our assumption that $\det \mathcal{C} \neq 0$ implies that this homogeneous algebraic system has only the solution $p\varphi = 0$; hence, $\mathcal{V}^-\varphi = 0$. But

$$\mathcal{V}_0^+\varphi = V_0\varphi = \mathcal{V}_0^-\varphi = 0,$$

so we also deduce that $\mathcal{V}^+\varphi = 0$, since $\mathcal{V}^+\varphi$ is the the unique solution of the homogeneous problem (D$^-$). Using Remark 2.17(iii), we now conclude that $\varphi = 0$. Thus, if the equation $V_0\varphi = 0$ has non-zero solutions, then $\det \mathcal{C} = 0$.

2.30. Corollary. *The null space of V_0 is the subspace of the null space of $W_0^* + \frac{1}{2}I$ consisting of all functions of the form $\varphi = \Phi k$, where k is any constant (3×1)-vector such that $\mathcal{C}k = 0$. More precisely, if $\operatorname{rank} \mathcal{C} = i$, $i = 0, 1, 2, 3$, then the dimension of the null space of V_0 is $3 - i$. In particular, if $\det \mathcal{C} \neq 0$, this space consists of the zero vector alone; if $\mathcal{C} = 0$, it coincides with the null space of $W_0^* + \frac{1}{2}I$.*

Proof. By Corollary 2.28, $V_0\varphi = 0$ implies that

$$\varphi = \Phi k, \quad \mathcal{C}k = 0 \tag{2.72}$$

for some constant $k \in \mathcal{M}_{3\times 1}$, so

$$(W_0^* + \tfrac{1}{2}I)\varphi = \big((W_0^* + \tfrac{1}{2}I)\Phi\big)k = 0;$$

that is, the null space of V_0 is the subspace of all $\varphi = \Phi k$ in the null space of $W_0^* + \frac{1}{2}I$ such that $\mathcal{C}k = 0$.

If $\operatorname{rank} \mathcal{C} = 0$, then $\mathcal{C} = 0$ and $V_0\Phi = 0$. This means that all the $\Phi^{(i)}$ are in the null space of V_0, so this space coincides with the span of $\{\Phi^{(i)}\}$, which is the null space of $W_0^* + \frac{1}{2}I$.

If $\operatorname{rank} \mathcal{C} = 1$, then only one of the columns of $\mathcal{C}$ is linearly independent, so there are two constant linearly independent (3×1)-vectors a and b such that

$$\mathcal{C}a = 0, \quad \mathcal{C}b = 0,$$

which implies that

$$V_0(\Phi a) = (V_0\Phi)a = (F\mathcal{C})a = F(\mathcal{C}a) = 0, \quad V_0(\Phi b) = 0.$$

This means that the null space of V_0 is spanned by the linearly independent vectors Φa and Φb; therefore, it is two-dimensional.

Similarly, if $\operatorname{rank} \mathcal{C} = 2$, then exactly two of the columns of $\mathcal{C}$ are linearly independent, so there is a constant non-zero $a \in \mathcal{M}_{3\times 1}$ such that $\mathcal{C}a = 0$. Hence,

$V_0(\Phi a) = 0$, which tells us that the null space of V_0 is the one-dimensional space spanned by Φa.

Finally, if rank $\mathcal{C} = 3$, then $\det \mathcal{C} \neq 0$ and the columns of $\mathcal{C}$ are linearly independent, so $\mathcal{C}k = 0$ in (2.72) implies that $k = 0$, which, in turn, yields $\varphi = 0$.

2.31. Example. From Example 2.27 it is obvious that for $R = e^{-1}$

$$\Phi = (2\pi)^{-1} e \begin{pmatrix} 1 & 0 & e^2 x_2 \\ 0 & 1 & -e^2 x_1 \end{pmatrix},$$

$$\mathcal{C} = (4\pi\mu)^{-1} e^2 \operatorname{diag}\{0, 0, 1\} \neq 0$$

and

$$V_0 \Phi = (4\pi\mu)^{-1} e^2 \begin{pmatrix} 0 & 0 & x_2 \\ 0 & 0 & -x_1 \end{pmatrix} = F\mathcal{C}.$$

Clearly, $\det \mathcal{C} = 0$, but $\mathcal{C} \neq 0$. As we know, the null space of $W_0^* + \frac{1}{2}I$ is the span of $\{\Phi^{(i)}\}$, which in this case coincides with the space $\mathcal{F}$ of rigid displacements, while the null space of V_0 is the span of $\{\Phi^{(1)}, \Phi^{(2)}, 0\}$, which is a proper subspace of $\mathcal{F}$.

2.32. Lemma. (i) *The characteristic matrix $\mathcal{C}$ is symmetric, that is,*

$$\mathcal{C}^T = \mathcal{C},$$

and if q is the operator defined on $C(\partial S)$ by

$$q\varphi = \int_{\partial S} \Phi^T \varphi \, ds,$$

then for any $\varphi \in C(\partial S)$

$$q(V_0 \varphi) = \mathcal{C}(p\varphi).$$

(ii) *For any $\psi \in C^{0,\alpha}(\partial S)$*

$$p(W_0^* \psi) = -\tfrac{1}{2} p\psi.$$

(iii) *For any $\psi \in C^{0,\alpha}(\partial S)$*

$$q(W_0 \psi) = -\tfrac{1}{2} q\psi.$$

(iv) *For any $\psi \in C^{1,\alpha}(\partial S)$*

$$p(N_0 \psi) = 0.$$

Proof. (i) Changing the order of integration and using (2.38), we find that

$$q(V_0\varphi) = \int_{\partial S} \Phi^{\mathrm{T}}(V_0\varphi)\,ds$$

$$= \int_{\partial S} \Phi^{\mathrm{T}} \left(\int_{\partial S} D(x,y)\varphi(y)\,ds(y) \right) ds(x)$$

$$= \int_{\partial S} \left(\int_{\partial S} D^{\mathrm{T}}(x,y)\Phi(x)\,ds(x) \right)^{\mathrm{T}} \varphi(y)\,ds(y)$$

$$= \int_{\partial S} \left(\int_{\partial S} D(y,x)\Phi(x)\,ds(x) \right)^{\mathrm{T}} \varphi(y)\,ds(y)$$

$$= \int_{\partial S} (V_0\Phi)^{\mathrm{T}} \varphi\,ds = \int_{\partial S} (FC)^{\mathrm{T}} \varphi\,ds$$

$$= \int_{\partial S} C^{\mathrm{T}} F^{\mathrm{T}} \varphi\,ds = C^{\mathrm{T}}(p\varphi). \tag{2.73}$$

If we now operate with q in $(2.67)_1$ and take (2.73) into account, we obtain

$$\mathcal{C}^{\mathrm{T}}(p\Phi) = q(V_0\Phi) = q(FC) = (qF)\mathcal{C} = (p\Phi)^{\mathrm{T}}\mathcal{C},$$

or $\mathcal{C}^{\mathrm{T}} = \mathcal{C}$. This brings (2.73) to the form $q(V_0\varphi) = \mathcal{C}(p\varphi)$.

(ii) Similarly, by Remark 2.17(v),

$$p(W_0^*\psi) = \int_{\partial S} F^{\mathrm{T}}(W_0^*\psi)\,ds$$

$$= \int_{\partial S} F^{\mathrm{T}}(x) \left(\int_{\partial S} T(\partial_x)D(x,y)\psi(y)\,ds(y) \right) ds(x)$$

$$= \int_{\partial S} F^{\mathrm{T}}(x) \left(\int_{\partial S} T(\partial_x)D(x,y)\psi(y)\,ds(y) \right) ds(x)$$

$$= \int_{\partial S} \left(\int_{\partial S} (T(\partial_x)D(x,y))^{\mathrm{T}} F(x)\,ds(x) \right)^{\mathrm{T}} \psi(y)\,ds(y)$$

$$= \int_{\partial S} \left(\int_{\partial S} P(y,x)F(x)\,ds(x) \right)^{\mathrm{T}} \psi(y)\,ds(y)$$

$$= \int_{\partial S} (W_0 F)^{\mathrm{T}} \psi\,ds = -\tfrac{1}{2} \int_{\partial S} F^{\mathrm{T}} \psi\,ds = -\tfrac{1}{2} p\psi.$$

(iii) By Theorem 2.22(ii),

$$\begin{aligned}
q(W_0\psi) &= \int_{\partial S} \Phi^{\mathrm{T}}(W_0\psi)\,ds = \int_{\partial S} \Phi^{\mathrm{T}}(x)\left(\int_{\partial S} P(x,y)\psi(y)\,ds(y)\right)ds(x) \\
&= \int_{\partial S} \Phi^{\mathrm{T}}(x)\left(\int_{\partial S} \bigl(T(\partial_y)D(y,x)\bigr)^{\mathrm{T}}\psi(y)\,ds(y)\right)ds(x) \\
&= \int_{\partial S}\left(\int_{\partial S} T(\partial_y)D(y,x)\Phi(x)\,ds(x)\right)^{\mathrm{T}}\psi(y)\,ds(y) \\
&= \int_{\partial S}(W_0^*\Phi)^{\mathrm{T}}\psi\,ds = -\tfrac{1}{2}\int_{\partial S}\Phi^{\mathrm{T}}\psi\,ds = -\tfrac{1}{2}q\psi.
\end{aligned}$$

(iv) Finally, by (2.19),

$$p(N_0\psi) = p\bigl(T(\mathcal{W}^+\psi)\bigr) = \int_{\partial S} F^{\mathrm{T}}T(\mathcal{W}^+\psi)\,ds = \int_{S^+} F^{\mathrm{T}}A(\mathcal{W}^+\varphi)\,da = 0.$$

2.7. The classical indirect method

We begin by considering the interior and exterior Dirichlet and Neumann problems. Thus, for $(D^\pm)$ and $(N^\pm)$ we seek solutions of the form

$$\begin{aligned}
u &= \mathcal{W}^+\varphi & &\text{for } (D^+), \\
u &= \mathcal{V}^+\psi & &\text{for } (N^+), \\
u &= \mathcal{W}^-\varphi + Fk & &\text{for } (D^-), \\
u &= \mathcal{V}^-\psi & &\text{for } (N^-),
\end{aligned}$$

where $k \in \mathcal{M}_{3\times 1}$ is constant and will be specified in the solution process.

Using the properties of the elastic potentials from Theorem 2.16 and the corresponding boundary conditions, we find that the unknown densities φ and ψ must satisfy the boundary integral equations

$$\begin{aligned}
(W_0 - \tfrac{1}{2}I)\varphi &= \mathcal{P} & &\text{for } (D^+), & &(\mathcal{D}_C^+) \\
(W_0^* + \tfrac{1}{2}I)\psi &= \mathcal{Q} & &\text{for } (N^+), & &(\mathcal{N}_C^+) \\
(W_0 + \tfrac{1}{2}I)\varphi &= \mathcal{R} - Fk & &\text{for } (D^-), & &(\mathcal{D}_C^-) \\
(W_0^* - \tfrac{1}{2}I)\psi &= \mathcal{S} & &\text{for } (N^-). & &(\mathcal{N}_C^-)
\end{aligned}$$

From the form of their kernels it is seen that $(\mathcal{D}_C^+)$, $(\mathcal{N}_C^-)$ and $(\mathcal{D}_C^-)$, $(\mathcal{N}_C^+)$ are pairwise adjoint. Also, as mentioned in Corollary 2.20, the Fredholm Alternative can be applied to them in the (real) dual system $(C^{0,\alpha}(\partial S), C^{0,\alpha}(\partial S))$, $\alpha \in (0,1)$, with the bilinear form

$$(\varphi, \psi) = \int_{\partial S} \varphi^T \psi \, ds.$$

2.33. Theorem. (i) $(\mathcal{D}_C^+)$ has a unique solution $\varphi \in C^{1,\alpha}(\partial S)$ for any prescribed $\mathcal{P} \in C^{1,\alpha}(\partial S)$.

Then (D^+) has the (unique) solution

$$u = \mathcal{W}^+ \varphi.$$

(ii) $(\mathcal{D}_C^-)$ with $k = q\mathcal{R}$ is solvable in $C^{1,\alpha}(\partial S)$ for any $\mathcal{R} \in C^{1,\alpha}(\partial S)$, and its solution is unique up to an arbitrary rigid displacement Fa.

Then (D^-) has the (unique) solution (in $\mathcal{A}^*$)

$$u = \mathcal{W}^- \varphi + F(q\mathcal{R}),$$

where φ is any solution of $(\mathcal{D}_C^-)$.

(iii) $(\mathcal{N}_C^+)$ is solvable in $C^{0,\alpha}(\partial S)$ for any $\mathcal{Q} \in C^{0,\alpha}(\partial S)$ such that $p\mathcal{Q} = 0$. In this case the solution is unique up to a term of the form Φa, where $a \in \mathcal{M}_{3\times 1}$ is constant and arbitrary.

Then (N^+) has the family of solutions

$$u = \mathcal{V}^+ \psi + Fk,$$

where ψ is any solution of $(\mathcal{N}_C^+)$ and $k \in \mathcal{M}_{3\times 1}$ is constant and arbitrary.

(iv) $(\mathcal{N}_C^-)$ has a unique solution $\psi \in C^{0,\alpha}(\partial S)$.

Then, if $p\mathcal{S} = 0$, (N^-) has the (unique) solution (in $\mathcal{A}$)

$$u = \mathcal{V}^- \psi.$$

Proof. (i), (iv) Theorem 2.22(i) states that the null spaces of $W_0^* - \frac{1}{2}I$ and $W_0 - \frac{1}{2}I$ consist only of the zero vector, which means that, by the Fredholm Alternative, equations $(\mathcal{D}_C^+)$ and $(\mathcal{N}_C^-)$ have unique solutions $\varphi, \psi \in C^{0,\alpha}(\partial S)$.

Given that $\mathcal{P} \in C^{1,\alpha}(\partial S)$, and taking Theorem 2.16(iv) into account, we see that the solution $\varphi = 2(W_0 \varphi + \mathcal{P})$ of $(\mathcal{D}_C^+)$ belongs in fact to $C^{1,\alpha}(\partial S)$; hence, $\mathcal{W}^+ \varphi \in C^{1,\alpha}(\bar{S}^+)$, and, since $A(\mathcal{W}^+ \varphi) = 0$, it follows that $u = \mathcal{W}^+ \varphi$ is the (unique) solution of (D^+).

Operating with p in $(\mathcal{N}_C^-)$ and using Lemma 2.32(ii), we conclude that

$$p\psi = -p\mathcal{S};$$

therefore, if $p\mathcal{S} = 0$, then $p\psi = 0$, which means that $\mathcal{V}^-\psi \in \mathcal{A}$. Also, $A(\mathcal{V}^-\psi) = 0$ and $\mathcal{V}^-\psi \in C^{1,\alpha}(\bar{S}^-)$ (see Theorem 2.16(ii)), so $u = \mathcal{V}^-\psi$ is the (unique) solution of (N$^-$).

(iii) By Theorem 2.22(ii), the null space of $W_0 + \frac{1}{2}I$ is spanned by $\{F^{(i)}\}$. Since

$$p\mathcal{Q} = \int_{\partial S} F^{\mathrm{T}} \mathcal{Q}\,ds = 0$$

is equivalent to

$$(F^{(i)}, \mathcal{Q}) = 0, \quad i = 1, 2, 3,$$

the Fredholm Alternative indicates that $(\mathcal{N}_C^+)$ is solvable in $C^{0,\alpha}(\partial S)$. We recall that the null space of $W_0^* + \frac{1}{2}I$ is spanned by $\{\Phi^{(i)}\}$; hence, the solution of $(\mathcal{N}_C^+)$ is unique up to a term Φa, where $a \in \mathcal{M}_{3\times 1}$ is constant and arbitrary.

For any solution $\psi \in C^{0,\alpha}(\partial S)$ of $(\mathcal{N}_C^+)$ and any constant $k \in \mathcal{M}_{3\times 1}$ we have $\mathcal{V}^+\psi + Fk \in C^{1,\alpha}(\bar{S}^+)$ and $A(\mathcal{V}^+\psi + Fk) = 0$, so $u = \mathcal{V}^+\psi + Fk$ is a solution of problem (N$^+$). The arbitrariness of k incorporates that arising from the term Φa in ψ, since

$$\mathcal{V}^+(\Phi a) = (\mathcal{V}^+\Phi)a = (F\mathcal{C})a = F(\mathcal{C}a).$$

(ii) For $k = q\mathcal{R} = \int_{\partial S} \Phi^{\mathrm{T}}\mathcal{R}\,ds$, (2.67)$_2$ yields

$$\int_{\partial S} \Phi^{\mathrm{T}}(\mathcal{R} - F(q\mathcal{R}))\,ds = q\mathcal{R} - \int_{\partial S}(\Phi^{\mathrm{T}}F)(q\mathcal{R})\,ds = q\mathcal{R} - \left(\int_{\partial S} F^{\mathrm{T}}\Phi\,ds\right)^{\mathrm{T}}(q\mathcal{R})$$

$$= q\mathcal{R} - (p\Phi)^{\mathrm{T}}(q\mathcal{R}) = 0,$$

which is equivalent to

$$(\Phi^{(i)}, \mathcal{R} - Fk) = (\Phi^{(i)}, \mathcal{R} - F(q\mathcal{R})) = 0, \quad i = 1, 2, 3.$$

By the Fredholm Alternative, this means that $(\mathcal{D}_C^-)$ is solvable in $C^{0,\alpha}(\partial S)$. Since the null space of $W_0 + \frac{1}{2}I$ is spanned by $\{F^{(i)}\}$, the solution of $(\mathcal{D}_C^-)$ is unique up to a term Fa, where $a \in \mathcal{M}_{3\times 1}$ is constant and arbitrary. Just as in the case of $(\mathcal{D}_C^+)$, φ actually belongs to $C^{1,\alpha}(\partial S)$, so $\mathcal{W}^-\varphi \in C^{1,\alpha}(\bar{S}^-)$. Also, Theorem 2.14(ii) shows that $\mathcal{W}^-\varphi + F(q\mathcal{R}) \in \mathcal{A}^*$, and $A(\mathcal{W}^-\varphi + F(q\mathcal{R})) = 0$. Hence, $u = \mathcal{W}^-\varphi + F(q\mathcal{R})$ is the (unique) solution of (D$^-$). The term Fa in φ is irrelevant, since $\mathcal{W}^-F = 0$.

For the Robin problems (R$^+$) and (R$^-$) we seek solutions of the form

$$u = \mathcal{V}^+\bigl(\varphi - \Phi(p\varphi)\bigr) + F(p\varphi), \tag{2.74}$$

$$u = \mathcal{V}^-\bigl(\varphi - \Phi(p\varphi)\bigr) + F(p\varphi). \tag{2.75}$$

Then the boundary conditions

$$Tu + \sigma u|_{\partial S} = \mathcal{K}, \quad Tu - \sigma u|_{\partial S} = \mathcal{L}$$

give rise, respectively, to the boundary integral equations

$$(W_0^* + \tfrac{1}{2}I)\bigl(\varphi - \Phi(p\varphi)\bigr) + \sigma V_0\bigl(\varphi - \Phi(p\varphi)\bigr) + \sigma F(p\varphi) = \mathcal{K}, \qquad (\mathcal{R}_C^+)$$

$$(W_0^* - \tfrac{1}{2}I)\bigl(\varphi - \Phi(p\varphi)\bigr) - \sigma V_0\bigl(\varphi - \Phi(p\varphi)\bigr) - \sigma F(p\varphi) = \mathcal{L}. \qquad (\mathcal{R}_C^-)$$

Since the dominant terms in the kernels of $(\mathcal{R}_C^+)$ and $(\mathcal{R}_C^-)$ are the same as in those of $(\mathcal{N}_C^+)$ and $(\mathcal{N}_C^-)$, the index of each of these equations is zero, so the Fredholm Alternative can be applied to them.

2.34. Theorem. *Let $\sigma \in C^{0,\alpha}(\partial S)$, $\alpha \in (0,1)$.*
(i) $(\mathcal{R}_C^+)$ has a unique solution $\varphi \in C^{0,\alpha}(\partial S)$ for any $\mathcal{K} \in C^{0,\alpha}(\partial S)$. Then the (unique) solution of (R$^+$) is given by (2.74).
(ii) $(\mathcal{R}_C^-)$ has a unique solution $\varphi \in C^{0,\alpha}(\partial S)$ for any $\mathcal{L} \in C^{0,\alpha}(\partial S)$. Then the (unique) solution of (R$^-$) is given by (2.75).

Proof. Since $\sigma \in C^{0,\alpha}(\partial S)$, the operators occurring in $(\mathcal{R}_C^+)$ and $(\mathcal{R}_C^-)$ map $C^{0,\alpha}(\partial S)$ to $C^{0,\alpha}(\partial S)$.

(i) Consider a solution $\bar\varphi$ of the homogeneous equation $(\mathcal{R}_C^+)$, in other words, a function $\bar\varphi \in C^{0,\alpha}(\partial S)$ such that

$$(W_0^* + \tfrac{1}{2}I)\bigl(\bar\varphi - \Phi(p\bar\varphi)\bigr) + \sigma V_0\bigl(\bar\varphi - \Phi(p\bar\varphi)\bigr) + \sigma F(p\bar\varphi) = 0. \tag{2.76}$$

This means that $\mathcal{V}^+\bigl(\bar\varphi - \Phi(p\bar\varphi)\bigr) + F(p\bar\varphi)$ is the (unique) solution of the homogeneous problem (R$^+$); therefore,

$$\mathcal{V}^+\bigl(\bar\varphi - \Phi(p\bar\varphi)\bigr) + F(p\bar\varphi) = 0. \tag{2.77}$$

Since $p\bigl(\bar\varphi - \Phi(p\bar\varphi)\bigr) = 0$, from (2.55), (2.76) and (2.77) we deduce that the function

$$\mathcal{U}^- = \mathcal{V}^-\bigl(\bar\varphi - \Phi(p\bar\varphi)\bigr) + F(p\bar\varphi)$$

satisfies

$$A\mathcal{U}^- = 0,$$
$$\mathcal{U}^-|_{\partial S} = V_0(\bar{\varphi} - \Phi(p\bar{\varphi})) + F(p\bar{\varphi}) = V_0^+(\bar{\varphi} - \Phi(p\bar{\varphi})) + F(p\bar{\varphi}) = 0,$$
$$\mathcal{U}^-(x) = \left[M^\infty p(\bar{\varphi} - \Phi(p\bar{\varphi})) + \mathcal{U}^{\mathcal{A}} + F(p\bar{\varphi})\right](x)$$
$$= (\mathcal{U}^{\mathcal{A}} + F(p\bar{\varphi}))(x) \quad \text{as } |x| \to \infty.$$

By Corollary 2.12(i),
$$F(p\bar{\varphi}) = 0, \quad \mathcal{V}^-(\bar{\varphi} - \Phi(p\bar{\varphi})) + F(p\bar{\varphi}) = \mathcal{V}^-(\bar{\varphi} - \Phi(p\bar{\varphi})) = 0.$$

Since the columns of F are linearly independent, this implies that $p\bar{\varphi} = 0$ and $\mathcal{V}^-\bar{\varphi} = 0$; also, (2.77) yields $\mathcal{V}^+\bar{\varphi} = 0$; hence, by Remark 2.17(iii), $\bar{\varphi} = 0$.

Since the homogeneous equation $(\mathcal{R}_C^+)$ has only the zero solution, the Fredholm Alternative states that $(\mathcal{R}_C^+)$ has a unique solution $\varphi \in C^{0,\alpha}(\partial S)$. By the properties of the elastic potentials in Theorem 2.16, u given by (2.74) belongs to $C^{1,\alpha}(\bar{S}^+)$ and satisfies $Au = 0$ (in S^+), so it is the (unique) solution of (R^+).

(ii) If $\bar{\varphi}$ is a solution of the homogeneous equation $(\mathcal{R}_C^-)$, that is,
$$(W_0^* - \tfrac{1}{2}I)(\bar{\varphi} - \Phi(p\bar{\varphi})) - \sigma V_0(\bar{\varphi} - \Phi(p\bar{\varphi})) - \sigma F(p\bar{\varphi}) = 0,$$

then, by (2.47), the function
$$\mathcal{U}^- = \mathcal{V}^-(\bar{\varphi} - \Phi(p\bar{\varphi})) + F(p\bar{\varphi})$$

satisfies
$$A\mathcal{U}^- = 0,$$
$$T\mathcal{U}^- - \sigma\mathcal{U}^-|_{\partial S} = 0,$$
$$\mathcal{U}^-(x) = \left[M^\infty p(\bar{\varphi} - \Phi(p\bar{\varphi})) + \mathcal{U}^{\mathcal{A}} + F(p\bar{\varphi})\right](x)$$
$$= (\mathcal{U}^{\mathcal{A}} + F(p\bar{\varphi}))(x) \quad \text{as } |x| \to \infty,$$

which is the homogeneous problem (R^-). Applying Corollary 2.12(i) once more, we conclude that
$$F(p\bar{\varphi}) = 0, \quad \mathcal{V}^-(\bar{\varphi} - \Phi(p\bar{\varphi})) + F(p\bar{\varphi}) = \mathcal{V}^-(\bar{\varphi} - \Phi(p\bar{\varphi})) = 0;$$

hence, as above, $p\bar{\varphi} = 0$ and
$$V_0^-\bar{\varphi} = 0 = V_0\bar{\varphi} = V_0^+\bar{\varphi},$$

so $\mathcal{V}^+\bar{\varphi} = 0$ as the unique solution of the homogeneous problem (D$^+$). Remark 2.17(iii) now implies that $\bar{\varphi} = 0$. Consequently, by the Fredholm Alternative, ($\mathcal{R}_C^-$) has a unique solution $\varphi \in C^{0,\alpha}(\partial S)$.

The function u given by (2.75) belongs to $C^{1,\alpha}(\bar{S}^-)$ and satisfies $Au = 0$ (in S^-). Also, $u \in \mathcal{A}^*$ since $p(\varphi - \Phi(p\varphi)) = 0$, which means that u is the (unique) solution of (R$^-$).

2.35. Remark. The sole purpose of the term $\Phi(p\varphi)$ in the density of $\mathcal{V}^\pm$ is to ensure that p applied to the density yields zero. This term can be replaced by any other that has the same effect. For example, in [21] the correction term in the density is $F(pF)^{-1}(p\varphi)$.

We can now make use of Theorem 2.33 to give the boundary curve ∂S another characterization, equivalent to that in Theorem 2.25, in terms of the solution of a generalized exterior Dirichlet problem.

2.36. Theorem. *For every simple closed C^2-curve ∂S and any $\alpha \in (0,1)$, there are a unique (2×3)-matrix $\bar{\Phi} \in C^\infty(S^-) \cap C^{1,\alpha}(\bar{S}^-)$ and a unique constant $\bar{C} \in \mathcal{M}_{3\times 3}$ such that*

$$A\bar{\Phi} = 0,$$
$$\bar{\Phi}|_{\partial S} = 0, \qquad (2.78)$$
$$\bar{\Phi}(x) = (M^\infty - F\bar{C} + \bar{\Phi}^{\mathcal{A}})(x) \quad as \ |x| \to \infty.$$

Proof. Since, by Lemma 2.15, $AM^\infty = 0$, the given problem is equivalent to finding $\bar{\Phi}$ such that

$$A(\bar{\Phi} - M^\infty) = 0,$$
$$(\bar{\Phi} - M^\infty)|_{\partial S} = -M^\infty|_{\partial S}, \qquad (2.79)$$
$$(\bar{\Phi} - M^\infty)(x) = (-F\bar{C} + \bar{\Phi}^{\mathcal{A}})(x) \quad as \ |x| \to \infty.$$

Regarding this as an exterior Dirichlet problem for each of the columns $(\bar{\Phi} - M^\infty)^{(i)}$, from Theorem 2.33(ii) with $\mathcal{R} = -(M^\infty)^{(i)}|_{\partial S}$ we deduce that (2.79) has a unique solution if

$$\bar{C} = q(M^\infty|_{\partial S}). \qquad (2.80)$$

By Theorem 2.33(ii), this solution can be expressed as

$$\bar{\Phi} - M^\infty = \mathcal{W}^-\hat{\bar{\Phi}} - Fq(M^\infty|_{\partial S}) = \mathcal{W}^-\hat{\bar{\Phi}} - F\bar{C}$$

with some (2×3)-matrix $\hat{\Phi} \in C^{1,\alpha}(\partial S)$. Then $\bar{\mathcal{C}}$ given by (2.80) and, from the above equality,

$$\bar{\Phi} = \mathcal{M}^\infty + \mathcal{W}^-\hat{\Phi} - F\bar{\mathcal{C}} = \mathcal{M}^\infty - F\bar{\mathcal{C}} + \bar{\Phi}^{\mathcal{A}}$$

are a solution pair for (2.78).

The difference $\tilde{\Phi}$, $\tilde{\mathcal{C}}$ of any two such solutions satisfies

$$A\tilde{\Phi} = 0,$$
$$\tilde{\Phi}|_{\partial S} = 0,$$
$$\tilde{\Phi}(x) = (-F\tilde{\mathcal{C}} + \tilde{\Phi}^{\mathcal{A}})(x) \quad \text{as } |x| \to \infty,$$

so, by Corollary 2.12(i),

$$F\tilde{\mathcal{C}} = 0, \quad \tilde{\Phi} = \tilde{\Phi}^{\mathcal{A}} = 0.$$

Since the columns of F are linearly independent, this implies that $\tilde{\mathcal{C}} = 0$; therefore, the solution pair of (2.78) is unique.

2.37. Theorem. *The pairs Φ, $\mathcal{C}$ and $\bar{\Phi}$, $\bar{\mathcal{C}}$ in Theorems 2.25 and 2.36 are connected by the relations*

$$\mathcal{C} = \bar{\mathcal{C}}, \quad \bar{\Phi} = \mathcal{V}^-\Phi - F\mathcal{C}, \quad \Phi = -T\bar{\Phi}.$$

Proof. By Theorem 2.25, $\mathcal{V}^-\Phi - F\mathcal{C}$ satisfies

$$A(\mathcal{V}^-\Phi - F\mathcal{C}) = 0,$$
$$\mathcal{V}_0^-\Phi - F\mathcal{C} = 0,$$
$$(\mathcal{V}^-\Phi - F\mathcal{C})(x) = \big(\mathcal{M}^\infty(p\Phi) + \mathcal{U}^{\mathcal{A}} - F\mathcal{C}\big)(x)$$
$$= (\mathcal{M}^\infty - F\mathcal{C} + \mathcal{U}^{\mathcal{A}})(x) \quad \text{as } |x| \to \infty,$$

which is problem (2.78). Since this problem has a unique solution, it follows that

$$\mathcal{C} = \bar{\mathcal{C}}, \quad \bar{\Phi} = \mathcal{V}^-\Phi - F\mathcal{C}. \tag{2.81}$$

Given that $(W_0^* - \tfrac{1}{2}I)\Phi = 0$, or $W_0^*\Phi = \tfrac{1}{2}\Phi$, and $TF = 0$, applying T to $(2.81)_2$ and taking $(2.49)_2$ into account, we obtain

$$T\bar{\Phi} = T(\mathcal{V}^-\Phi) = (W_0^* - \tfrac{1}{2}I)\Phi = \tfrac{1}{2}\Phi - \tfrac{1}{2}\Phi = -\Phi.$$

2.8. The alternative indirect method

As in the case of the Laplace equation, in plane elasticity we can also devise a procedure that allows us to replace the singular integral equations of the Dirichlet problems arising in the classical indirect method with weakly singular ones.

Thus, seeking the solution of (D$^+$) in the form $u = \mathcal{V}^+\varphi$, we are led to the boundary integral equation

$$V_0\varphi = \mathcal{P}. \qquad (\mathcal{D}_{\mathrm{A}}^+)$$

2.38. Theorem. *If* $\det \mathcal{C} \neq 0$, *then* $(\mathcal{D}_{\mathrm{A}}^+)$ *has a unique solution* $\varphi \in C^{0,\alpha}(\partial S)$ *for any* $\mathcal{P} \in C^{1,\alpha}(\partial S)$, $\alpha \in (0,1)$.

In this case,

$$u = \mathcal{V}^+\varphi$$

is the (unique) solution of (D$^+$).

Proof. Operating with N_0 in $(\mathcal{D}_{\mathrm{A}}^+)$ and applying the composition formula $(2.56)_2$, we see that any solution of $(\mathcal{D}_{\mathrm{A}}^+)$ also satisfies

$$(W_0^{*2} - \tfrac{1}{4}I)\varphi = N_0\mathcal{P}. \qquad (2.82)$$

By Theorem 2.24(i), the null space of the adjoint operator $W^2 - \tfrac{1}{4}I$ is spanned by the rigid displacements $\{F^{(i)}\}$. Lemma 2.32(iv) now yields

$$\int_{\partial S} F^{\mathrm{T}}(N_0\mathcal{P})\,ds = p(N_0\mathcal{P}) = 0,$$

which is equivalent to

$$(F^{(i)}, N_0\mathcal{P}) = 0, \quad i = 1, 2, 3,$$

so, by the Fredholm Alternative, (2.82) is solvable in $C^{0,\alpha}(\partial S)$ and its solutions are of the form

$$\varphi = \bar\varphi + \Phi a,$$

where $\bar\varphi$ is any (fixed) solution and $a \in \mathcal{M}_{3\times 1}$ is constant and arbitrary. Writing

$$(W_0^{*2} - \tfrac{1}{4}I)(\bar\varphi + \Phi a) - N_0\mathcal{P} = 0,$$

operating with V_0 in this equation and taking the composition formulae $(2.56)_1$ and $(2.57)_2$ into account, we find that

$$(W_0^2 - \tfrac{1}{4}I)(V_0(\bar\varphi + \Phi a) - \mathcal{P}) = 0,$$

which, by Theorem 2.24(i), means that

$$V_0(\bar{\varphi} + \Phi a) - \mathcal{P} = Fa',$$

where $a' \in \mathcal{M}_{3\times 1}$ is constant. Recalling that $V_0\Phi = F\mathcal{C}$, we deduce that

$$V_0\bar{\varphi} - \mathcal{P} = Fa' - F(\mathcal{C}a) = Fa'', \tag{2.83}$$

where a'' is a constant (3×1)-vector that depends on the choice of $\bar{\varphi}$. Operating with q in (2.83) and applying Lemma 2.32(i), we find that

$$\mathcal{C}(p\bar{\varphi}) - q\mathcal{P} = (qF)a'' = (p\Phi)^\mathrm{T} a'' = a''. \tag{2.84}$$

Next, since $\mathcal{C}$ is invertible and

$$V_0\big(\bar{\varphi} - \Phi(\mathcal{C}^{-1}a'')\big) = V_0\bar{\varphi} - (F\mathcal{C})(\mathcal{C}^{-1}a'') = V_0\bar{\varphi} - Fa'' = \mathcal{P},$$

we use (2.84) to conclude that

$$\varphi = \bar{\varphi} - \Phi(\mathcal{C}^{-1}a'') = \bar{\varphi} + \Phi\big(\mathcal{C}^{-1}(q\mathcal{P}) - p\bar{\varphi}\big) \tag{2.85}$$

is a solution of $(\mathcal{D}_\mathrm{A}^+)$ in $C^{0,\alpha}(\partial S)$.

The difference $\tilde{\varphi}$ of two solutions satisfies $V_0\tilde{\varphi} = 0$, which, by Theorem 2.29, yields $\tilde{\varphi} = 0$; in other words, $(\mathcal{D}_\mathrm{A}^+)$ has a unique solution regardless of the choice of $\bar{\varphi}$.

By Theorem 2.16, the function $u = \mathcal{V}^+\varphi$ belongs to $C^{1,\alpha}(\bar{S}^+)$ and satisfies $A(\mathcal{V}^+\varphi) = 0$, so it is the (unique) solution of (D$^+$). By (2.85), we may also write it as

$$u = \mathcal{V}^+\varphi = \mathcal{V}^+\bar{\varphi} + F\big(q\mathcal{P} - \mathcal{C}(p\bar{\varphi})\big). \tag{2.86}$$

If $\det \mathcal{C} = 0$, then the above argument breaks down because $\mathcal{C}^{-1}$ does not exist. Instead, here we seek the solution as

$$u = \mathcal{V}^+\varphi + Fc,$$

where $c \in \mathcal{M}_{3\times 1}$ is constant, so the boundary integral equation is

$$V_0\varphi = \mathcal{P} - Fc. \tag{$\tilde{\mathcal{D}}_\mathrm{A}^+$}$$

Operating with q in this equation and applying Lemma 2.32(i), we find that

$$\mathcal{C}(p\varphi) = q(V_0\varphi) = q\mathcal{P} - (qF)c = q\mathcal{P} - c;$$

therefore,
$$c = q\mathcal{P} - \mathcal{C}(p\varphi). \tag{2.87}$$

2.39. Theorem. *If* $\det \mathcal{C} = 0$, *then* $(\tilde{\mathcal{D}}_A^+)$ *with c given by (2.87) is solvable in* $C^{0,\alpha}(\partial S)$ *for any* $\mathcal{P} \in C^{1,\alpha}(\partial S)$, $\alpha \in (0,1)$, *and its solution is unique up to a term of the form* Φa, *where* $a \in \mathcal{M}_{3\times 1}$ *is constant and arbitrary.*

In this case the (unique) solution of (D^+) *is again given by (2.86), where $\bar{\varphi}$ is any solution of* $(\tilde{\mathcal{D}}_A^+)$.

Proof. Replacing c from (2.87) in $(\tilde{\mathcal{D}}_A^+)$, we see that the latter can be written in the form
$$V_0\varphi - (FC)(p\varphi) = \mathcal{P} - F(q\mathcal{P}). \tag{2.88}$$

Proceeding as in the proof of Theorem 2.38 and recalling that $N_0 F = 0$, we deduce that any solution of (2.88) is also a solution of (2.82), which is solvable in $C^{0,\alpha}(\partial S)$, and we arrive again at the equality
$$V_0\bar{\varphi} - \mathcal{P} = Fa' - F(\mathcal{C}a), \tag{2.89}$$

where $\bar{\varphi}$ is any (fixed) solution. Operating with q on both sides and making use of Lemma 2.32(i), we obtain
$$\mathcal{C}(p\bar{\varphi}) - q\mathcal{P} = a' - \mathcal{C}a,$$

so
$$a' = \mathcal{C}(p\bar{\varphi}) - q\mathcal{P} + \mathcal{C}a;$$

hence, (2.89) becomes
$$V_0\bar{\varphi} = \mathcal{P} + (FC)(p\bar{\varphi}) - F(q\mathcal{P}),$$

which is equation (2.88). This means that, conversely, any solution of (2.82) is also a solution of (2.88), and, therefore, of $(\tilde{\mathcal{D}}_A^+)$. As already mentioned, the arbitrariness in the solution of (2.82) is Φa; consequently, the solution of $(\tilde{\mathcal{D}}_A^+)$ has the same arbitrariness.

The function
$$u = \mathcal{V}^+\bar{\varphi} + Fc = \mathcal{V}^+\bar{\varphi} + F(q\mathcal{P} - \mathcal{C}(p\bar{\varphi}))$$

belongs to $C^{1,\alpha}(\bar{S}^+)$ and satisfies $Au = 0$ (in S^+), so it is the (unique) solution of problem (D^+). Once again, its uniqueness is not affected by the choice of the solution $\bar{\varphi}$ of $(\tilde{\mathcal{D}}_A^+)$.

2.9. The modified indirect method

As in Chapter 1, we can still have a unique solution for $(\mathcal{D}_A^+)$ whatever the value of $\det \mathcal{C}$ if we choose a suitably modified matrix of fundamental solutions for $-A$. In this case we take

$$D^H(x,y) = D(x,y) + F(x)HF^{\mathrm{T}}(y), \tag{2.90}$$

where $H \in \mathcal{M}_{3\times 3}$ is constant and symmetric but otherwise arbitrary, and introduce the corresponding modified single layer potential

$$(V^H\varphi)(x) = \int_{\partial S} D^H(x,y)\varphi(y)\,ds(y) = (V\varphi)(x) + F(x)H(p\varphi). \tag{2.91}$$

Since $TF = 0$, from (2.90) it is clear that the modified double layer potential coincides with the original one, so we continue to use the old symbols for the operators W_0, W_0^* and N_0.

Throughout this section we use the notation $\mathcal{V}^{H\pm}$, $\mathcal{V}_0^{H\pm}$ and V_0^H with the obvious meaning.

2.40. Theorem. *If* $\det(H+\mathcal{C}) \neq 0$, $\varphi \in C^{0,\alpha}(\partial S)$, $\alpha \in (0,1)$, *and* $V_0^H\varphi = 0$, *then* $\varphi = 0$.

Proof. Let

$$\mathcal{U}^- = \mathcal{V}^{H-}\varphi - (\mathcal{V}^-\Phi)(p\varphi) + (F\mathcal{C})(p\varphi) = \mathcal{V}^-\bigl(\varphi - \Phi(p\varphi)\bigr) + F(H+\mathcal{C})(p\varphi).$$

Since $AF = 0$ and $V_0^H\varphi = 0$, the function $\mathcal{U}^-$ satisfies

$$A\mathcal{U}^- = 0,$$

$$\mathcal{U}^-|_{\partial S} = V_0^H\varphi - (V_0\Phi)(p\varphi) + (F\mathcal{C})(p\varphi) = V_0^H\varphi = 0,$$

$$\begin{aligned}
\mathcal{U}^-(x) &= \bigl[\mathcal{V}^-\bigl(\varphi - \Phi(p\varphi)\bigr) + F(H+\mathcal{C})(p\varphi)\bigr](x) \\
&= \bigl[M^\infty p\bigl(\varphi - \Phi(p\varphi)\bigr) + \mathcal{U}^\mathcal{A} + F(H+\mathcal{C})(p\varphi)\bigr](x) \\
&= \bigl[\mathcal{U}^\mathcal{A} + F(H+\mathcal{C})(p\varphi)\bigr](x)) \quad \text{as } |x| \to \infty.
\end{aligned}$$

This is the homogeneous problem (D^-), and Corollary 2.12(i) implies that

$$F(H+\mathcal{C})(p\varphi) = 0,$$
$$\mathcal{U}^- = \mathcal{V}^-\bigl(\varphi - \Phi(p\varphi)\bigr) + F(H+\mathcal{C})(p\varphi) = \mathcal{V}^-\bigl(\varphi - \Phi(p\varphi)\bigr) = 0. \tag{2.92}$$

Since the columns of the matrix F are linearly independent, from $(2.92)_1$ it follows that $(H+C)(p\varphi) = 0$, which, given that $\det(H+C) \neq 0$, yields $p\varphi = 0$; therefore, equation $(2.92)_2$ reduces to $\mathcal{V}^-\varphi = 0$. In turn, this means that

$$\mathcal{V}_0^+ \varphi = V_0\varphi = \mathcal{V}_0^-\varphi = 0;$$

therefore, $\mathcal{V}^+\varphi = 0$ as the unique solution of the homogeneous problem (D$^+$). Using Remark 2.17(iii), we now conclude that $\varphi = 0$.

2.41. Theorem. *Compositions* (2.56) *and* (2.57) *hold if V_0 is replaced by V_0^H.*

Proof. Substituting

$$V_0\varphi = V_0^H \varphi - (FH)(p\varphi)$$

in (2.56) and (2.57)$_2$ and working with $\varphi \in C^{0,\alpha}(\partial S)$ and $\psi \in C^{1,\alpha}(\partial S)$, we arrive at the equalities

$$W_0\bigl(V_0^H \varphi - (FH)(p\varphi)\bigr) = V_0^H(W_0^* \varphi) - (FH)p(W_0^* \varphi),$$
$$N_0\bigl(V_0^H \varphi - (FH)(p\varphi)\bigr) = (W_0^{*2} - \tfrac{1}{4}I)\varphi, \qquad (2.93)$$
$$V_0^H(N_0\psi) - (FH)p(N_0\psi) = (W_0^2 - \tfrac{1}{4}I)\psi.$$

From Remark 2.17(v), Lemma 2.32(ii), Theorem 2.24(ii) and Lemma 2.32(iv), respectively, we see that

$$W_0\bigl((FH)(p\varphi)\bigr) = (W_0 F)H(p\varphi) = -\tfrac{1}{2}(FH)(p\varphi),$$
$$(FH)p(W_0^* \varphi) = -\tfrac{1}{2}(FH)(p\varphi),$$
$$N_0\bigl((FH)(p\varphi)\bigr) = (N_0 F)H(p\varphi) = 0,$$
$$(FH)p(N_0\psi) = 0,$$

so the proof follows immediately from (2.93).

For simplicity, in this section we continue to quote the composition formulae as (2.56) and (2.57), but we understand them with V_0^H instead of V_0.

We now seek the solution of (D$^+$) in the form $u = \mathcal{V}^{H+}\varphi$ with H such that $\det(H+C) \neq 0$ chosen a priori. Then the problem reduces to the solution of the Fredholm equation of the first kind

$$V_0^H \varphi = \mathcal{P}. \qquad (\mathcal{D}_M^+)$$

2.42. Theorem. $(\mathcal{D}_M^+)$ has a unique solution $\varphi \in C^{0,\alpha}(\partial S)$ for any $\mathcal{P} \in C^{1,\alpha}(\partial S)$, $\alpha \in (0,1)$.

In this case,
$$u = \mathcal{V}^{H+}\varphi$$
is the (unique) solution of (D^+).

Proof. We follow the procedure set out in the proof of Theorem 2.38.

Operating with N_0 in $(\mathcal{D}_M^+)$, we deduce that any solution of $(\mathcal{D}_M^+)$ also satisfies
$$(W_0^{*2} - \tfrac{1}{4}I)\varphi = N_0\mathcal{P}. \tag{2.94}$$

The null space of the operator of the adjoint equation is spanned by $\{F^{(i)}\}$ and, by Lemma 2.32(iv),
$$\int_{\partial S} F^T(N_0\mathcal{P})\, ds = p(N_0\mathcal{P}) = 0,$$
which is equivalent to
$$(F^{(i)}, N_0\mathcal{P}) = 0, \quad i = 1, 2, 3.$$

Hence, by the Fredholm Alternative, (2.94) is solvable in $C^{0,\alpha}(\partial S)$ and its solutions are
$$\varphi = \bar{\varphi} + \Phi a,$$
where $\bar{\varphi}$ is any (fixed) solution and $a \in \mathcal{M}_{3\times 1}$ is constant and arbitrary. Writing
$$(W_0^{*2} - \tfrac{1}{4}I)(\bar{\varphi} + \Phi a) - N_0\mathcal{P} = 0,$$
applying V_0 to this equation and taking the composition formulae $(2.56)_1$ and $(2.57)_2$ into account, we obtain
$$(W_0^2 - \tfrac{1}{4}I)\bigl(V_0^H(\bar{\varphi} + \Phi a) - \mathcal{P}\bigr) = 0.$$

By Theorem 2.24(i), this means that
$$V_0^H(\bar{\varphi} + \Phi a) - \mathcal{P} = Fa',$$
where $a' \in \mathcal{M}_{3\times 1}$ is constant. The above equation can be rewritten in the form
$$V_0^H\bar{\varphi} - \mathcal{P} = Fa' - V_0^H(\Phi a) = Fa' - \bigl(V_0\Phi + (FH)(p\Phi)\bigr)a$$
$$= F\bigl[a' - (H+C)a\bigr] = Fa'', \tag{2.95}$$

or, equivalently,
$$V_0\bar\varphi + (FH)(p\bar\varphi) - \mathcal{P} = Fa''.$$

Operating with q in this equality, we find that, by Lemma 2.32(i),
$$(H + \mathcal{C})(p\bar\varphi) - q\mathcal{P} = a''. \tag{2.96}$$

The condition $\det(H + \mathcal{C}) \neq 0$ implies that $H + \mathcal{C}$ is invertible, so, by (2.95),
$$\begin{aligned}
V_0^H\big(\bar\varphi - \Phi(H+\mathcal{C})^{-1}a''\big) &= V_0^H\bar\varphi - (V_0^H\Phi)(H+\mathcal{C})^{-1}a'' \\
&= V_0^H\bar\varphi - \big[V_0\Phi + (FH)(p\Phi)\big](H+\mathcal{C})^{-1}a'' \\
&= V_0^H\bar\varphi - F(H+\mathcal{C})(H+\mathcal{C})^{-1}a'' = V_0^H\bar\varphi - Fa'' = \mathcal{P}.
\end{aligned}$$

Consequently,
$$\varphi = \bar\varphi - \Phi(H+\mathcal{C})^{-1}a'' \in C^{0,\alpha}(\partial S),$$

with a'' given by (2.96), is a solution of $(\mathcal{D}_M^+)$.

This solution is unique since the difference $\tilde\varphi$ of two solutions satisfies $V_0^H\tilde\varphi = 0$, which, by Theorem 2.40, yields $\tilde\varphi = 0$.

The function
$$\begin{aligned}
u = \mathcal{V}^{H+}\varphi &= \mathcal{V}^{H+}\big(\bar\varphi - \Phi(H+\mathcal{C})^{-1}a''\big) \\
&= \mathcal{V}^{H+}\big(\bar\varphi - \Phi(p\bar\varphi) - (H+\mathcal{C})^{-1}(q\mathcal{P})\big)
\end{aligned}$$

belongs to $C^{1,\alpha}(\bar S^+)$ and satisfies $Au = 0$ (in S^+); therefore, u is the (unique) solution of (D^+).

2.43. Remark. It is useful to find out the connection between two different representations of the solution u of (D^+). Let $H_1 \neq H_2$ be two distinct constant symmetric (3×3)-matrices such that $\det(H_\alpha + \mathcal{C}) \neq 0$, and let $\varphi^{(\alpha)}$ be the densities corresponding to them. Constructing u in terms of each of these densities and using the boundary condition, we obtain
$$V_0^{H_1}\varphi^{(1)} = V_0^{H_2}\varphi^{(2)} = \mathcal{P},$$

or
$$V_0\varphi^{(1)} + (FH_1)(p\varphi^{(1)}) = V_0\varphi^{(2)} + (FH_2)(p\varphi^{(2)}) = \mathcal{P}.$$

This implies that
$$V_0(\varphi^{(1)} - \varphi^{(2)}) = F\big(H_2(p\varphi^{(2)}) - H_1(p\varphi^{(1)})\big),$$

which, by Corollary 2.28, yields

$$\varphi^{(1)} - \varphi^{(2)} = \Phi k, \quad H_2(p\varphi^{(2)}) - H_1(p\varphi^{(1)}) = \mathcal{C} k \qquad (2.97)$$

for some constant $k \in \mathcal{M}_{3\times 1}$; hence,

$$p\varphi^{(1)} - p\varphi^{(2)} = (p\Phi)k = k. \qquad (2.98)$$

From $(2.97)_2$ we deduce that

$$\mathcal{C}(p\varphi^{(1)} - p\varphi^{(2)}) = H_2(p\varphi^{(2)}) - H_1(p\varphi^{(1)}),$$

or

$$(H_1 + \mathcal{C})(p\varphi^{(1)}) = (H_2 + \mathcal{C})(p\varphi^{(2)}).$$

The condition $\det(H_\alpha + \mathcal{C}) \neq 0$ implies that $H_2 + \mathcal{C}$ is invertible, so

$$p\varphi^{(2)} = (H_2 + \mathcal{C})^{-1}(H_1 + \mathcal{C})(p\varphi^{(1)}).$$

We now rewrite (2.98) as

$$k = \left[E_3 - (H_2 + \mathcal{C})^{-1}(H_1 + \mathcal{C}) \right](p\varphi^{(1)})$$

and conclude from $(2.97)_1$ that

$$\varphi^{(2)} = \varphi^{(1)} + \Phi\left[(H_2 + \mathcal{C})^{-1}(H_1 + \mathcal{C}) - E_3 \right](p\varphi^{(1)}).$$

This formula shows how any solution of $(\mathcal{D}_M^+)$ can be constructed from the solution corresponding to a particular choice of matrix H.

As in the case of the Laplace equation, this indirect method makes it possible for us to solve a generalized exterior Dirichlet problem. Specifically, we consider the boundary value problem

$$\begin{aligned} (Au)(x) &= 0, \quad x \in S^-, \\ u(x) &= \mathcal{R}(x), \quad x \in \partial S, \\ u(x) &= (M^\infty s + u^{\mathcal{A}^*})(x) \quad \text{as } |x| \to \infty, \end{aligned} \qquad (\mathrm{D}_G^-)$$

where s is a given constant (3×1)-vector. When $s = 0$, (D_G^-) becomes (D^-).

100 PLANE STRAIN

2.44. Theorem. (D_G^-) has a unique solution for any $\mathcal{R} \in C^{1,\alpha}(\partial S)$, $\alpha \in (0,1)$, which can be expressed as a modified single layer potential.

Proof. In Theorem 2.42 we have shown that we can find $\varphi \in C^{0,\alpha}(\partial S)$ such that

$$\mathcal{V}_0^{H+}\varphi = V_0^H\varphi = \mathcal{R}. \tag{2.99}$$

Then $\mathcal{V}^{H-}\varphi$ satisfies the first two equations in (D_G^-). But there is no guarantee that the third one also holds for this function. Consequently, we modify the procedure and seek the solution in the form

$$u = \mathcal{V}^{L-}\psi, \quad \psi = \varphi + \Phi k, \tag{2.100}$$

where φ is the density mentioned in (2.99), L is a matrix of the same type as H and the constant (3×1)-vector k is determined in the construction. Clearly, $u \in C^{1,\alpha}(\bar{S}^-)$ and $Au = 0$ (in S^-).

By (2.47), as $|x| \to \infty$

$$(\mathcal{V}^{L-}\psi)(x) = \left[\mathcal{V}^{L-}\varphi + (\mathcal{V}^{L-}\Phi)k\right](x)$$
$$= M^\infty(p\varphi) + (FL)(p\varphi) + M^\infty(p\Phi)k + (FL)(p\Phi)k + \mathcal{U}^{\mathcal{A}}$$
$$= M^\infty(p\varphi + k) + \mathcal{U}^{\mathcal{A}^*}.$$

This matches the far-field pattern of (D_G^-) if

$$k = s - p\varphi. \tag{2.101}$$

We also need to check if this function satisfies the boundary condition. Since

$$\mathcal{V}^{L-}\varphi = \mathcal{V}^-\varphi + (FL)(p\varphi)$$
$$= \mathcal{V}^-\varphi + (FH)(p\varphi) + F(L-H)(p\varphi) = \mathcal{V}^{H-}\varphi + F(L-H)(p\varphi),$$

from (2.101) and (2.100) we see that

$$\mathcal{V}_0^{L-}\psi = \mathcal{V}_0^{L-}(\varphi + \Phi k) = V_0^L\varphi + (V_0^L\Phi)(s - p\varphi)$$
$$= V_0^H\varphi + F(L-H)(p\varphi) + \bigl(V_0\Phi + (FL)(p\Phi)\bigr)(s - p\varphi)$$
$$= \mathcal{R} + F\bigl((L-H)(p\varphi) + (L+\mathcal{C})(s - p\varphi)\bigr). \tag{2.102}$$

When $s \neq 0$, we have $\mathcal{V}_0^{L-}\psi = \mathcal{R}$ if

$$(L-H)(p\varphi) + (L+\mathcal{C})(s - p\varphi) = 0,$$

or
$$Ls = H(p\varphi) + C(p\varphi - s) = (H + C)(p\varphi) - Cs. \qquad (2.103)$$

The matrix L is determined from this equation, but with some arbitrariness. However, this arbitrariness is unimportant for the solution of (D_G^-) since, by $(2.100)_1$ and (2.103),

$$\begin{aligned}
u &= \mathcal{V}^{L-}\psi = \mathcal{V}^{L-}(\varphi + \Phi k) \\
&= \mathcal{V}^{L-}\bigl(\varphi + \Phi(s - p\varphi)\bigr) \\
&= \mathcal{V}^-\bigl(\varphi + \Phi(s - p\varphi)\bigr) + (FL)p\bigl(\varphi + \Phi(s - p\varphi)\bigr) \\
&= \mathcal{V}^-\bigl(\varphi + \Phi(s - p\varphi)\bigr) + F(Ls) \\
&= \mathcal{V}^-\bigl(\varphi + \Phi(s - p\varphi)\bigr) + F\bigl[(H + C)(p\varphi) - Cs\bigr].
\end{aligned}$$

If $s = 0$, that is, in the case of (D^-), (2.101) reduces to

$$k = -p\varphi$$

and (2.102) becomes

$$\mathcal{V}_0^{L-}\psi = \mathcal{R} - F(H + C)(p\varphi),$$

which is independent of L. To satisfy the boundary condition we need to have $(H + C)(p\varphi) = 0$, but $\det(H + C) \neq 0$, and we cannot guarantee that $p\varphi = 0$ in general. But we see that

$$\begin{aligned}
V_0^H\bigl(\varphi - \Phi(p\varphi)\bigr) &+ F(H + C)(p\varphi) \\
&= \mathcal{R} - \bigl(V_0\Phi + (FH)(p\Phi)\bigr)(p\varphi) + F(H + C)(p\varphi) \\
&= \mathcal{R} - (FC)(p\varphi) - (FH)(p\varphi) + F(H + C)(p\varphi) = \mathcal{R},
\end{aligned}$$

and, since $p\bigl(\varphi - \Phi(p\varphi)\bigr) = 0$,

$$\mathcal{V}^{H-}\bigl(\varphi - \Phi(p\varphi)\bigr) + F(H + C)(p\varphi) \in \mathcal{A}^*.$$

Consequently, the (unique) solution of (D^-) is

$$u = \mathcal{V}^{H-}\bigl(\varphi - \Phi(p\varphi)\bigr) + F(H + C)(p\varphi),$$

where φ is the density in (2.99).

2.10. The refined indirect method

To bypass the need to know $\mathcal{C}$ beforehand in order to choose a suitable matrix H in the modified indirect technique, here we seek the solution of (D^+) as

$$u = \mathcal{V}^+ \varphi - Fc, \qquad (2.104)$$

where $c \in \mathcal{M}_{3\times 1}$ is constant. Since this expression contains two unknown quantities φ and c, we adjoin an additional condition on the former, namely, $p\varphi = s$, with a constant $s \in \mathcal{M}_{3\times 1}$ chosen a priori. Applying the boundary condition, we then obtain the system of integral equations

$$V_0 \varphi - Fc = \mathcal{P}, \quad p\varphi = s. \qquad (\mathcal{D}_\mathrm{R}^+)$$

2.45. Theorem. $(\mathcal{D}_\mathrm{R}^+)$ *has a unique solution* φ, c *with* $\varphi \in C^{0,\alpha}(\partial S)$ *for any* $\mathcal{P} \in C^{1,\alpha}(\partial S)$, $\alpha \in (0,1)$.

Then the (unique) solution of (D^+) *is given by* (2.104).

Proof. We operate with N_0 in the first equation in $(\mathcal{D}_\mathrm{R}^+)$ and use the composition formula $(2.56)_2$ and Theorem 2.24(ii) to deduce that any solution of $(\mathcal{D}_\mathrm{R}^+)$ also satisfies the Fredholm equation of the second kind

$$(W_0^{*2} - \tfrac{1}{4}I)\varphi = N_0 \mathcal{P}. \qquad (2.105)$$

Applying Lemma 2.32(iv), we find that

$$p(N_0 \mathcal{P}) = 0,$$

which is equivalent to

$$(F^{(i)}, N_0 \mathcal{P}) = 0, \quad i = 1, 2, 3,$$

so, by the Fredholm Alternative, (2.105) is solvable in $C^{0,\alpha}(\partial S)$ and its solution is unique up to a solution of the corresponding homogeneous equation; that is,

$$\varphi = \bar{\varphi} + \Phi a, \qquad (2.106)$$

where $\bar{\varphi}$ is any (fixed) solution and $a \in \mathcal{M}_{3\times 1}$ is constant and arbitrary. Now we can write

$$(W_0^{*2} - \tfrac{1}{4}I)(\bar{\varphi} + \Phi a) = N_0 \mathcal{P},$$

which, after operation by V_0 and use of the compositions $(2.56)_1$ and $(2.57)_2$, yields

$$(W_0^2 - \tfrac{1}{4}I)(V_0(\bar\varphi + \Phi a) - \mathcal{P}) = 0.$$

By Theorems 2.24(i) and 2.22(ii), this means that

$$V_0(\bar\varphi + \Phi a) - \mathcal{P} = Fc, \tag{2.107}$$

where $c \in \mathcal{M}_{3\times 1}$ is constant; therefore, we have a solution pair for the first equation in $(\mathcal{D}_R^+)$. Now we need to specify a so that the second equation is also satisfied.

Applying p in (2.106), we obtain

$$p\varphi = p(\bar\varphi + \Phi a) = p\bar\varphi + a;$$

hence, the second equation in $(\mathcal{D}_R^+)$ holds if

$$a = s - p\bar\varphi.$$

Also, operating with q in (2.107), we find that, by Lemma 2.32(i),

$$\mathcal{C}p(\bar\varphi + \Phi a) - q\mathcal{P} = (qF)c = (p\Phi)^\mathrm{T} c = c,$$

or

$$c = \mathcal{C}(p\bar\varphi + a) - q\mathcal{P} = \mathcal{C}(p\bar\varphi + s - p\bar\varphi) - q\mathcal{P} = \mathcal{C}s - q\mathcal{P}.$$

Consequently, a solution of $(\mathcal{D}_R^+)$ is

$$\varphi = \bar\varphi + \Phi(s - p\bar\varphi), \quad c = \mathcal{C}s - q\mathcal{P}. \tag{2.108}$$

The difference φ, c of two solutions satisfies

$$V_0\varphi - Fc = 0, \quad p\varphi = 0.$$

The later equality implies that $\mathcal{V}^-\varphi \in \mathcal{A}$. Hence, $\mathcal{V}^-\varphi - Fc$ is a solution of the homogeneous problem (D^-), so from Corollary 2.12(i) it follows that

$$Fc = 0, \quad \mathcal{V}^-\varphi = 0.$$

Thus,

$$\mathcal{V}_0^+\varphi = V_0\varphi = \mathcal{V}_0^-\varphi = 0,$$

so $\mathcal{V}^+\varphi = 0$, because $\mathcal{V}^+\varphi$ is the unique solution of the homogeneous problem (D$^+$). By Remark 2.17(iii), $\varphi = 0$. At the same time, the linear independence of the columns of F leads to $c = 0$, which completes the proof of uniqueness.

Since u given by (2.104) belongs to $C^{1,\alpha}(\bar{S}^+)$ and $Au = 0$ (in S^+), this function is the (unique) solution of (D$^+$).

2.46. Remark. Clearly, the representation $u = \mathcal{V}^+\varphi - Fc$ is not unique. Let $\varphi^{(\alpha)}$, $c^{(\alpha)}$ be two solutions of $(\mathcal{D}_\mathrm{R}^+)$, corresponding to $s^{(1)} \neq s^{(2)}$. Then (2.108) yields
$$\varphi^{(2)} = \varphi^{(1)} + \Phi(s^{(2)} - s^{(1)}), \quad c^{(2)} = c^{(1)} + \mathcal{C}(s^{(2)} - s^{(1)}). \tag{2.109}$$

The generalized exterior problem (D$_\mathrm{G}^-$) in §2.9 can also be solved by this method.

2.47. Theorem. (D$_\mathrm{G}^-$) *has a unique solution for any* $\mathcal{R} \in C^{1,\alpha}(\partial S)$, $\alpha \in (0,1)$, *which is constructed from the same density and additive rigid displacement as in* (D$^+$) *with* $\mathcal{P}$ *replaced by* $\mathcal{R}$, *and with s taken from the far-field pattern of the problem.*

Proof. Let φ, c be the solution of $(\mathcal{D}_\mathrm{R}^+)$ with boundary data $\mathcal{R}$. Then, by (2.47),
$$A(\mathcal{V}^-\varphi - Fc) = 0,$$
$$\mathcal{V}_0^-\varphi - Fc = V_0\varphi - Fc = \mathcal{R},$$
$$(\mathcal{V}^-\varphi - Fc)(x) = \big(M^\infty(p\varphi) + \mathcal{U}^\mathcal{A} - Fc\big)(x) = (M^\infty s + \mathcal{U}^{\mathcal{A}^*})(x) \quad \text{as } |x| \to \infty;$$
that is, $\mathcal{V}^-\varphi - Fc$ is a solution of (D$_\mathrm{G}^-$), which is unique since the difference of two solutions satisfies the homogeneous problem (D$^-$).

2.48. Remarks. (i) As in Chapter 1, this method provides the solutions of (D$^+$) and (D$_\mathrm{G}^-$) by means of the same formula.

(ii) We can now compute the characteristic matrices $\mathcal{C}$ and Φ.

Let $s^{(0)}$ and $s^{(i)}$, $i = 1,2,3$, be any set of 4 constant (3×1)-vectors such that $s^{(i)} - s^{(0)}$ are linearly independent. Then the constant (3×3)-matrix S with columns $s^{(i)} - s^{(0)}$ is invertible. We construct the solutions $\varphi^{(0)}$, $c^{(0)}$ and $\varphi^{(i)}$, $c^{(i)}$ of (D$^+$) corresponding to the vectors $s^{(0)}$ and $s^{(i)}$ and the same arbitrary data $\mathcal{P} \in C^{1,\alpha}(\partial S)$, and define the (3×3)-matrices Ψ and C whose columns are, respectively, $\varphi^{(i)} - \varphi^{(0)}$ and $c^{(i)} - c^{(0)}$. By (2.109), we can now write
$$\Phi S = \Psi, \quad \mathcal{C}S = C,$$
from which we obtain
$$\Phi = S^{-1}\Psi, \quad \mathcal{C} = S^{-1}C.$$

2.11. The direct method

We now set up our boundary integral equations for (D$^\pm$), (N$^\pm$) and (R$^\pm$) starting from the Somigliana representation formulae.

Let u be a solution of the equation $(Au)(x) = 0$, $x \in S^+$. By $(2.52)_1$ restricted to the boundary ∂S,

$$V_0(Tu) - (W_0 + \tfrac{1}{2}I)(u|_{\partial S}) = 0.$$

In the Dirichlet problem (D$^+$) we know $u|_{\partial S} = \mathcal{P}$, while $Tu = \varphi$ is unknown; in the Neumann problem (N$^+$), $Tu = \mathcal{Q}$ is known and $u|_{\partial S} = \psi$ is unknown. Thus, the corresponding boundary integral equations for (D$^+$) and (N$^+$) are, respectively,

$$V_0 \varphi = (W_0 + \tfrac{1}{2}I)\mathcal{P}, \tag{$\mathcal{D}_\mathrm{D}^+$}$$

$$(W_0 + \tfrac{1}{2}I)\psi = V_0 \mathcal{Q}. \tag{$\mathcal{N}_\mathrm{D}^+$}$$

Formula $(2.53)_1$ for a function u that satisfies $(Au)(x) = 0$, $x \in S^-$, is applicable only if $u \in \mathcal{A}$, which is indeed the case in (N$^-$). Restricting this formula to ∂S, we arrive at

$$-V_0(Tu) + (W_0 - \tfrac{1}{2}I)(u|_{\partial S}) = 0,$$

so the exterior Neumann problem gives rise to the equation

$$(W_0 - \tfrac{1}{2}I)\psi = V_0 \mathcal{S}. \tag{$\mathcal{N}_\mathrm{D}^-$}$$

The solution $u \in \mathcal{A}^*$ of (D$^-$) requires an alternative representation formula, which can easily be derived from the usual one. Writing $u = u^\mathcal{A} + Fc$, where $c \in \mathcal{M}_{3\times 1}$ is constant, we apply $(2.53)_1$ to $u^\mathcal{A} = u - Fc \in \mathcal{A}$ and find that

$$u - Fc = -\mathcal{V}^-\bigl(T(u - Fc)\bigr) + \mathcal{W}^-\bigl((u - Fc)|_{\partial S}\bigr).$$

Since $TF = 0$ and, by Remark 2.17(v), $\mathcal{W}^- F = 0$, we obtain the representation

$$u = -\mathcal{V}^-(Tu) + \mathcal{W}^-(u|_{\partial S}) + Fc. \tag{2.110}$$

The restriction of this equality to ∂S has the form

$$-V_0(Tu) + (W_0 - \tfrac{1}{2}I)(u|_{\partial S}) + Fc = 0,$$

which leads to the boundary integral equation

$$V_0 \varphi = (W_0 - \tfrac{1}{2}I)\mathcal{R} + Fc. \tag{$\mathcal{D}_\mathrm{D}^-$}$$

The rigid displacement Fc is found during the construction of the solution.

It is clear that $(\mathcal{D}_D^+)$ and $(\mathcal{D}_D^-)$ are equations of the first kind, so they have to be considered by themselves. The equations for $(N^\pm)$ are investigated in the next section.

The unknown function φ in $(\mathcal{D}_D^+)$ and $(\mathcal{D}_D^-)$ satisfies an additional condition because it is the vector of the moments and transverse shear force on ∂S, that is, $\varphi = Tu$. Using (2.19) in Lemma 2.3, we immediately see that

$$p\varphi = 0. \tag{2.111}$$

In the exterior Dirichlet problem, (2.111) also guarantees that the solution u belongs to $\mathcal{A}^*$ (see Theorem 2.14(i)), as stipulated in Definition 2.4.

2.49. Theorem. (i) *For any $\mathcal{P} \in C^{1,\alpha}(\partial S)$, $\alpha \in (0,1)$, the pair of equations $(\mathcal{D}_D^+)$ and (2.111) has a unique solution $\varphi \in C^{0,\alpha}(\partial S)$.*

Then (D^+) has the (unique) solution

$$u = \mathcal{V}^+ \varphi - \mathcal{W}^+ \mathcal{P}. \tag{2.112}$$

(ii) *For any $\mathcal{R} \in C^{1,\alpha}(\partial S)$, $\alpha \in (0,1)$, the pair of equations $(\mathcal{D}_D^-)$ with $c = q\mathcal{R}$ and (2.111) has a unique solution $\varphi \in C^{0,\alpha}(\partial S)$.*

Then (D^-) has the (unique) solution

$$u = -\mathcal{V}^- \varphi + \mathcal{W}^- \mathcal{R} + F(q\mathcal{R}). \tag{2.113}$$

Proof. (i) We operate with N_0 in $(\mathcal{D}_D^+)$ and use the compositions $(2.56)_2$ and $(2.57)_1$ to deduce that any solution of $(\mathcal{D}_D^+)$ is also a solution of the equation of the second kind

$$(W_0^{*2} - \tfrac{1}{4}I)\varphi = (W_0^* + \tfrac{1}{2}I)(N_0\mathcal{P}). \tag{2.114}$$

By Lemma 2.32(ii),

$$\int_{\partial S} F^T(W_0^* + \tfrac{1}{2}I)(N_0\mathcal{P})\, ds = p\big(W_0^*(N_0\mathcal{P})\big) + \tfrac{1}{2}p(N_0\mathcal{P})$$

$$= -\tfrac{1}{2}p(N_0\mathcal{P}) + \tfrac{1}{2}p(N_0\mathcal{P}) = 0.$$

Since this is equivalent to

$$\big(F^{(i)}, (W_0^* + \tfrac{1}{2}I)(N_0\mathcal{P})\big) = 0, \quad i = 1, 2, 3,$$

it follows that, by the Fredholm Alternative, equation (2.114) is solvable in $C^{0,\alpha}(\partial S)$ and its solutions can be written as

$$\varphi = \bar{\varphi} + \Phi a, \qquad (2.115)$$

where $\bar{\varphi}$ is any (fixed) solution and $a \in \mathcal{M}_{3\times 1}$ is constant and arbitrary. Writing

$$(W_0^{*2} - \tfrac{1}{4}I)(\bar{\varphi} + \Phi a) = (W_0^* + \tfrac{1}{2}I)(N_0 \mathcal{P})$$

and applying V_0 to both sides, we see that, by $(2.56)_1$ and $(2.57)_2$,

$$(W_0^2 - \tfrac{1}{4}I)\bigl[V_0(\bar{\varphi} + \Phi a) - (W_0 + \tfrac{1}{2}I)\mathcal{P}\bigr] = 0,$$

which, in view of Theorem 2.24(i), yields

$$V_0(\bar{\varphi} + \Phi a) - (W_0 + \tfrac{1}{2}I)\mathcal{P} = Fa', \qquad (2.116)$$

or

$$V_0\bar{\varphi} - (W_0 + \tfrac{1}{2}I)\mathcal{P} = Fa' - F(\mathcal{C}a), \qquad (2.117)$$

where $a' \in \mathcal{M}_{3\times 1}$ is constant and arbitrary.

Using p in (2.115) leads to

$$p\varphi = p\bar{\varphi} + (p\Phi)a = p\bar{\varphi} + a,$$

so the solution in (2.115) that satisfies (2.111) corresponds to $a = -p\bar{\varphi}$; that is,

$$\varphi = \bar{\varphi} - \Phi(p\bar{\varphi}). \qquad (2.118)$$

To check if this is a solution of $(\mathcal{D}_D^+)$, we operate with q in (2.117) for $a = -p\bar{\varphi}$. By Lemma 2.32(i),(iii),

$$\mathcal{C}(p\bar{\varphi}) + \tfrac{1}{2}q\mathcal{P} - \tfrac{1}{2}q\mathcal{P} = a' + \mathcal{C}(p\bar{\varphi}),$$

or $a' = 0$, so (2.116) takes the form

$$V_0\bigl(\bar{\varphi} - \Phi(p\bar{\varphi})\bigr) = (W_0 + \tfrac{1}{2}I)\mathcal{P},$$

which is equation $(\mathcal{D}_D^+)$. This shows that for any solution $\bar{\varphi}$ of (2.114), the function (2.118) satisfies both $(\mathcal{D}_D^+)$ and (2.111).

The difference $\tilde{\varphi}$ of two solutions of this pair of equations satisfies

$$V_0\tilde{\varphi} = 0, \quad p\tilde{\varphi} = 0.$$

From Corollary 2.30 it follows that $\tilde{\varphi} = \Phi k$ for some constant $k \in \mathcal{M}_{3\times 1}$. Then

$$0 = p\tilde{\varphi} = (p\Phi)k = k,$$

so $\tilde{\varphi} = 0$, which means that the solution (2.118) of $(\mathcal{D}_D^+)$ and (2.111) is unique, regardless of the choice of the solution $\bar{\varphi}$ of (2.114) used in its construction.

Now we must check that u given by (2.112) is the solution of (D^+). First of all, $\varphi \in C^{0,\alpha}(\partial S)$ and $\mathcal{P} \in C^{1,\alpha}(\partial S)$ imply that $\mathcal{V}^+\varphi, \mathcal{W}^+\mathcal{P} \in C^{1,\alpha}(\bar{S}^+)$; hence, $u \in C^{1,\alpha}(\bar{S}^+)$. Also, $Au = 0$ (in S^+). To verify the boundary condition, we restrict (2.112) to ∂S; then, by $(\mathcal{D}_D^+)$,

$$u|_{\partial S} = V_0\varphi - (W_0 - \tfrac{1}{2}I)\mathcal{P} = \left[V_0\varphi - (W_0 + \tfrac{1}{2}I)\mathcal{P}\right] + \mathcal{P} = \mathcal{P},$$

as required.

(ii) We manipulate $(\mathcal{D}_D^-)$ as in (i) and, by Theorem 2.24(ii), deduce that any solution of $(\mathcal{D}_D^-)$ is also a solution of the equation of the second kind

$$(W_0^{*2} - \tfrac{1}{4}I)\varphi = (W_0^* - \tfrac{1}{2}I)(N_0\mathcal{R}). \tag{2.119}$$

This equation is solvable in $C^{0,\alpha}(\partial S)$ since, by Lemma 2.32(ii),(iv),

$$\int_{\partial S} F^{\mathrm{T}}(W_0^* - \tfrac{1}{2}I)(N_0\mathcal{R})\,ds = p\big(W_0^*(N_0\mathcal{R})\big) - \tfrac{1}{2}p(N_0\mathcal{R})$$
$$= -\tfrac{1}{2}p(N_0\mathcal{R}) - \tfrac{1}{2}p(N_0\mathcal{R}) = -p(N_0\mathcal{R}) = 0.$$

Given that the null space of $W_0^{*2} - \tfrac{1}{4}I$ is spanned by $\{\Phi^{(i)}\}$, the solutions of (2.119) are of the form (2.115), where $\bar{\varphi}$ is any (fixed) solution of (2.119).

Continuing to follow the procedure used in (i), we find that

$$V_0(\bar{\varphi} + \Phi a) - (W_0 - \tfrac{1}{2}I)\mathcal{R} = Fa', \tag{2.120}$$

or

$$V_0\bar{\varphi} - (W_0 - \tfrac{1}{2}I)\mathcal{R} = Fa' - F(\mathcal{C}a) \tag{2.121}$$

with a constant and arbitrary $a' \in \mathcal{M}_{3\times 1}$, and that the solution (2.115) of (2.119) which also satisfies (2.111) is once again (2.118), corresponding to $a = -p\bar{\varphi}$.

We verify that (2.118) is a solution of ($\mathcal{D}_D^-$) by operating with q in (2.121) for $a = -p\bar{\varphi}$. In view of Lemma 2.32(i),(iii), this yields

$$\mathcal{C}(p\bar{\varphi}) + \tfrac{1}{2}q\mathcal{R} + \tfrac{1}{2}q\mathcal{R} = a' + \mathcal{C}(p\bar{\varphi}),$$

or $a' = q\mathcal{R}$. This makes (2.120) coincide with ($\mathcal{D}_D^-$) for $c = q\mathcal{R}$. The uniqueness of the solution (2.118) of ($\mathcal{D}_D^-$) and (2.111), irrespective of the choice of the solution $\bar{\varphi}$ of (2.119) in its construction, is shown as in (i).

We check that u given by (2.113) is the solution of (D$^-$). The restriction of (2.113) to ∂S and ($\mathcal{D}_D^-$) with $c = q\mathcal{R}$ lead to

$$\begin{aligned} u|_{\partial S} &= -V_0\varphi + (W_0 + \tfrac{1}{2}I)\mathcal{R} + F(q\mathcal{R}) \\ &= \left[-V_0\varphi + (W_0 - \tfrac{1}{2}I)\mathcal{R} + F(q\mathcal{R})\right] + \mathcal{R} = \mathcal{R}. \end{aligned}$$

Also, $u \in C^{1,\alpha}(\bar{S}^-)$, $Au = 0$ (in S^-) and, since φ satisfies (2.111), $u \in \mathcal{A}^*$.

We now discuss the Robin problems, and begin by writing the boundary condition for (R$^+$) in the form

$$Tu = -\sigma u|_{\partial S} + \mathcal{K}.$$

We replace this in the representation formula (2.52)$_1$ and obtain

$$u = -\mathcal{V}^+(\sigma u|_{\partial S}) - \mathcal{W}^+(u|_{\partial S}) + \mathcal{V}^+\mathcal{K}, \qquad (2.122)$$

which, restricted to ∂S, yields

$$u|_{\partial S} = -V_0(\sigma u|_{\partial S}) - (W_0 - \tfrac{1}{2}I)(u|_{\partial S}) + V_0\mathcal{K}.$$

With the notation $\varphi = u|_{\partial S}$, this can be rewritten as

$$V_0(\sigma\varphi) + (W_0 + \tfrac{1}{2}I)\varphi = V_0\mathcal{K}. \qquad (\mathcal{R}_D^+)$$

We seek the solution of (R$^-$) in the form

$$u = u^{\mathcal{A}} + Fc,$$

where $c \in \mathcal{M}_{3\times 1}$ is constant. Using (2.110) with

$$Tu = \sigma u|_{\partial S} + \mathcal{L},$$

we find that

$$u = -\mathcal{V}^-(\sigma u|_{\partial S} + \mathcal{L}) + \mathcal{W}^-(u|_{\partial S}) + F c. \tag{2.123}$$

Replacing $u|_{\partial S} = \varphi$ in (2.123) and restricting this equality to ∂S, we arrive at the boundary integral equation

$$V_0(\sigma\varphi) - (W_0 - \tfrac{1}{2}I)\varphi - Fc = -V_0\mathcal{L}, \tag{$\mathcal{R}_D^-$}$$

with c to be determined during the process of solution.

As in the Dirichlet problems, in $(\mathrm{R}^\pm)$ the density φ must satisfy an additional condition. Since $\varphi = u|_{\partial S}$, we write the boundary conditions of (R^+) and (R^-), respectively, as

$$Tu = -\sigma\varphi + \mathcal{K},$$
$$Tu = \sigma\varphi + \mathcal{L}.$$

Operating with p in both these equalities and recalling that, by Lemma 2.3, we have $p(Tu) = 0$, we arrive at the constraints

$$p(\sigma\varphi - \mathcal{K}) = 0, \tag{2.124}$$
$$p(\sigma\varphi + \mathcal{L}) = 0. \tag{2.125}$$

2.50. Theorem. *Suppose that $\sigma \in C^{0,\alpha}(\partial S)$ and that $\det \mathcal{C} \neq 0$.*

(i) *The pair of equations $(\mathcal{R}_D^+)$ and (2.124) has a unique solution $\varphi \in C^{1,\alpha}(\partial S)$ for any $\mathcal{K} \in C^{0,\alpha}(\partial S)$.*

Then (R^+) has the (unique) solution

$$u = -\mathcal{V}^+(\sigma\varphi) - \mathcal{W}^+\varphi + \mathcal{V}^+\mathcal{K}.$$

(ii) *The pair of equations $(\mathcal{R}_D^-)$ with $c = q\varphi$ and (2.125) has a unique solution $\varphi \in C^{1,\alpha}(\partial S)$ for any $\mathcal{L} \in C^{0,\alpha}(\partial S)$.*

Then (R^-) has the (unique) solution

$$u = -\mathcal{V}^-(\sigma\varphi) + \mathcal{W}^-\varphi + F(q\varphi) - \mathcal{V}^-\mathcal{L}.$$

Proof. (i) Consider a solution $\bar{\varphi}$ of the homogeneous equation $(\mathcal{R}_D^+)$; in other words, a function $\bar{\varphi}$ such that

$$V_0(\sigma\bar{\varphi}) + (W_0 + \tfrac{1}{2}I)\bar{\varphi} = 0, \tag{2.126}$$

or

$$V_0(\sigma\bar{\varphi}) + (W_0 - \tfrac{1}{2}I)\bar{\varphi} = -\bar{\varphi}. \tag{2.127}$$

THE DIRECT METHOD   111

This means that, by (2.126) and (2.47), the function

$$\mathcal{U}^- = \mathcal{V}^-(\sigma\bar{\varphi}) + \mathcal{W}^-\bar{\varphi} - (\mathcal{V}^-\Phi)p(\sigma\bar{\varphi}) \tag{2.128}$$

satisfies

$$A\mathcal{U}^- = 0,$$
$$\mathcal{U}^-|_{\partial S} = V_0(\sigma\bar{\varphi}) + (W_0 + \tfrac{1}{2}I)\bar{\varphi} - (V_0\Phi)p(\sigma\bar{\varphi}) = -F\big(\mathcal{C}p(\sigma\bar{\varphi})\big),$$
$$\mathcal{U}^-(x) = \big[M^\infty p(\sigma\bar{\varphi}) - M^\infty(p\Phi)p(\sigma\bar{\varphi}) + \mathcal{U}^{\mathcal{A}}\big](x) = \mathcal{U}^{\mathcal{A}}(x) \quad \text{as } |x| \to \infty,$$

which is an exterior Dirichlet problem. By Corollary 2.12(ii),

$$F\big(\mathcal{C}p(\sigma\bar{\varphi})\big) = 0, \quad \mathcal{U}^- = 0; \tag{2.129}$$

hence, since the columns of F are linearly independent and $\det\mathcal{C} \neq 0$, we first obtain $\mathcal{C}p(\sigma\bar{\varphi}) = 0$ and then $p(\sigma\bar{\varphi}) = 0$. Now combining this with (2.128) and the second equality (2.129), we deduce that

$$\mathcal{V}^-(\sigma\bar{\varphi}) + \mathcal{W}^-\bar{\varphi} = 0,$$

which, under operation by T, leads to

$$(W_0^* - \tfrac{1}{2}I)(\sigma\bar{\varphi}) + N_0\bar{\varphi} = 0, \tag{2.130}$$

or

$$(W_0^* + \tfrac{1}{2}I)(\sigma\bar{\varphi}) + N_0\bar{\varphi} = \sigma\bar{\varphi}. \tag{2.131}$$

By (2.131) and (2.127), the function

$$\mathcal{U}^+ = \mathcal{V}^+(\sigma\bar{\varphi}) + \mathcal{W}^+\bar{\varphi}$$

satisfies

$$A\mathcal{U}^+ = 0,$$
$$T\mathcal{U}^+ + \sigma\mathcal{U}^+|_{\partial S}$$
$$= \big[(W_0^* + \tfrac{1}{2}I)(\sigma\bar{\varphi}) + N_0\bar{\varphi}\big] + \sigma\big[V_0(\sigma\bar{\varphi}) + (W_0 - \tfrac{1}{2}I)\bar{\varphi}\big] = \sigma\bar{\varphi} - \sigma\bar{\varphi} = 0.$$

Clearly, the unique solution of this homogeneous problem (R$^+$) is

$$\mathcal{U}^+ = \mathcal{V}^+(\sigma\bar{\varphi}) + \mathcal{W}^+\bar{\varphi} = 0.$$

Applying T to the second equality, we obtain

$$(W_0^* + \tfrac{1}{2}I)(\sigma\bar{\varphi}) + N_0\bar{\varphi} = 0. \tag{2.132}$$

We now subtract (2.130) from (2.132) and arrive at $\sigma\bar{\varphi} = 0$, which implies that $\bar{\varphi} = 0$. Consequently, since the homogeneous equation $(\mathcal{R}_D^+)$ has only the zero solution, it follows that, by the Fredholm Alternative, $(\mathcal{R}_D^+)$ itself has a unique solution $\varphi \in C^{0,\alpha}(\partial S)$. From the mapping properties of the boundary operators (Theorem 2.16) we see that, in fact,

$$\varphi = 2[V_0\mathcal{K} - V_0(\sigma\varphi) - W_0\varphi] \in C^{1,\alpha}(\partial S).$$

We need to check that this solution also satisfies (2.124). Operating with q in $(\mathcal{R}_D^+)$ and using Lemma 2.32(i),(iii), we see that

$$\mathcal{C}p(\sigma\varphi) - \tfrac{1}{2}q\varphi + \tfrac{1}{2}q\varphi = \mathcal{C}(p\mathcal{K}),$$

or

$$\mathcal{C}p(\sigma\varphi - \mathcal{K}) = 0.$$

Since $\det \mathcal{C} \neq 0$, we conclude that $p(\sigma\varphi - \mathcal{K}) = 0$, as required.

We now verify that $u = -\mathcal{V}^+(\sigma\varphi) - \mathcal{W}^+\varphi + \mathcal{V}^+\mathcal{K}$ is the solution of (R^+). By Theorem 2.16, $\varphi, \sigma, \mathcal{K} \in C^{0,\alpha}(\partial S)$ implies that $u \in C^{1,\alpha}(\bar{S}^+)$. In addition, $Au = 0$ (in S^+). Restricting u to ∂S, computing Tu and using $(\mathcal{R}_D^+)$, we obtain

$$Tu + \sigma u|_{\partial S} = (W_0^* + \tfrac{1}{2}I)(\mathcal{K} - \sigma\varphi) - N_0\varphi + \sigma[V_0(\mathcal{K} - \sigma\varphi) - (W_0 - \tfrac{1}{2}I)\varphi]$$
$$= (W_0^* + \tfrac{1}{2}I)(\mathcal{K} - \sigma\varphi) - N_0\varphi + \sigma\varphi.$$

By $(2.56)_1$ and $(2.57)_2$, the function

$$\mathcal{U} = [(W_0^* + \tfrac{1}{2}I)(\mathcal{K} - \sigma\varphi) - N_0\varphi + \sigma\varphi] - \mathcal{K}$$
$$= (W_0^* - \tfrac{1}{2}I)(\mathcal{K} - \sigma\varphi) - N_0\varphi$$

satisfies

$$V_0\mathcal{U} = (V_0(W_0^* - \tfrac{1}{2}I))(\mathcal{K} - \sigma\varphi) - (V_0 N_0)\varphi$$
$$= (W_0 - \tfrac{1}{2}I)V_0(\mathcal{K} - \sigma\varphi) - (W_0^2 - \tfrac{1}{4}I)\varphi$$
$$= (W_0 - \tfrac{1}{2}I)[V_0(\mathcal{K} - \sigma\varphi) - (W_0 + \tfrac{1}{2}I)\varphi] = 0.$$

Since $\det \mathcal{C} \neq 0$, from Theorem 2.29 it follows that $\mathcal{U} = 0$. The definition of $\mathcal{U}$ now shows that $Tu + \sigma u|_{\partial S} = 0$.

(ii) Operating with q in $(\mathcal{R}_D^-)$ for $c = q\varphi$ and taking Lemma 2.32(i),(iii) into account yields

$$\mathcal{C}p(\sigma\varphi) + \tfrac{1}{2}q\varphi + \tfrac{1}{2}q\varphi - q\varphi = -\mathcal{C}(p\mathcal{L}),$$

or

$$\mathcal{C}p(\sigma\varphi + \mathcal{L}) = 0.$$

As in (i), we again infer that $p(\sigma\varphi + \mathcal{L}) = 0$, which means that any solution of $(\mathcal{R}_D^-)$ also satisfies (2.125).

Consider a solution $\bar{\varphi}$ of the homogeneous equation $(\mathcal{R}_D^-)$ with $c = q\varphi$, that is, a function $\bar{\varphi}$ such that

$$V_0(\sigma\bar{\varphi}) - (W_0 - \tfrac{1}{2}I)\bar{\varphi} - F(q\bar{\varphi}) = 0, \qquad (2.133)$$

also written as

$$V_0(\sigma\bar{\varphi}) - (W_0 + \tfrac{1}{2}I)\bar{\varphi} - F(q\bar{\varphi}) = -\bar{\varphi}. \qquad (2.134)$$

Then, by (2.133), the function

$$\mathcal{U}^+ = \mathcal{V}^+(\sigma\bar{\varphi}) - \mathcal{W}^+\bar{\varphi} - F(q\bar{\varphi})$$

satisfies

$$A\mathcal{U}^+ = 0,$$

$$\mathcal{U}^+|_{\partial S} = V_0(\sigma\bar{\varphi}) - (W_0 - \tfrac{1}{2}I)\bar{\varphi} - F(q\bar{\varphi}) = 0,$$

which is the homogeneous problem (D^+); therefore,

$$\mathcal{U}^+ = \mathcal{V}^+(\varphi\bar{\varphi}) - \mathcal{W}^+\bar{\varphi} - F(q\bar{\varphi}) = 0.$$

Since $TF = 0$, applying T to the second equality leads to

$$(W_0^* + \tfrac{1}{2}I)(\sigma\bar{\varphi}) - N_0\bar{\varphi} = 0, \qquad (2.135)$$

or

$$(W_0^* - \tfrac{1}{2}I)(\sigma\bar{\varphi}) - N_0\bar{\varphi} = -\sigma\bar{\varphi}. \qquad (2.136)$$

On the other hand, for $\mathcal{L} = 0$ condition (2.125) reduces to $p(\sigma\bar{\varphi}) = 0$, so from (2.136) and (2.134) it follows that the function

$$\mathcal{U}^- = \mathcal{V}^-(\sigma\bar{\varphi}) - \mathcal{W}^-\bar{\varphi} - F(q\bar{\varphi})$$

satisfies

$$A\mathcal{U}^- = 0,$$

$$T\mathcal{U}^- - \sigma\mathcal{U}^-|_{\partial S} = \left[(W_0^* - \tfrac{1}{2}I)(\sigma\bar\varphi) - N_0\bar\varphi\right] - \sigma\left[\mathcal{V}_0(\sigma\bar\varphi) - (W_0 + \tfrac{1}{2}I)\bar\varphi - F(q\bar\varphi)\right]$$
$$= -\sigma\bar\varphi - \sigma(-\bar\varphi) = 0,$$

$$\mathcal{U}^-(x) = \left[M^\infty p(\sigma\bar\varphi) + \mathcal{U}^{\mathcal{A}} - F(q\bar\varphi)\right](x) = \left[\mathcal{U}^{\mathcal{A}} - F(q\bar\varphi)\right](x) \quad \text{as } |x| \to \infty.$$

This is the homogeneous problem (R$^-$), so, by Corollary 2.12(i),

$$F(q\bar\varphi) = 0, \quad \mathcal{U}^- = \mathcal{V}^-(\sigma\bar\varphi) - \mathcal{W}^-\bar\varphi - F(q\bar\varphi) = \mathcal{V}^-(\sigma\bar\varphi) - \mathcal{W}^-\bar\varphi = 0.$$

Applying T to the last equality above yields

$$(W_0^* - \tfrac{1}{2}I)(\sigma\bar\varphi) - N_0\bar\varphi = 0. \tag{2.137}$$

If we subtract (2.137) from (2.135), then as in (i) we obtain $\bar\varphi = 0$. Thus, $(\mathcal{R}_\text{D}^-)$ has a unique solution $\varphi \in C^{0,\alpha}(\partial S)$ which satisfies (2.125) and which, by the mapping properties of the boundary operators, belongs actually to $C^{1,\alpha}(\partial S)$.

The fact that $u = -\mathcal{V}^-(\sigma\varphi) + \mathcal{W}^-\varphi + F(q\varphi) - \mathcal{V}^-\mathcal{L}$ is the solution of (R$^-$) is shown as in (i), but this time we make use of $(\mathcal{R}_\text{D}^-)$ with $c = q\varphi$, and (2.125) implies that $u \in \mathcal{A}^*$, as required.

2.51. Remarks. (i) If $\det \mathcal{C} = 0$, then we need to use a modified fundamental solution of the form proposed for the Dirichlet problems in §2.9.

(ii) Theorem 2.50 holds even when $\det \mathcal{C} = 0$ if σ is a constant matrix.

2.12. The substitute direct method

Equations of the first kind are not very helpful sometimes because they may give rise to ill-posed problems. If instead we want to deal with equations of the second kind in (D$^\pm$), then we apply T to the Somigliana formulae (2.52)$_1$ and (2.110) and use Theorem 2.16 to find that

$$Tu = (W_0^* + \tfrac{1}{2}I)(Tu) - N_0(u|_{\partial S}),$$
$$Tu = -(W_0^* - \tfrac{1}{2}I)(Tu) + N_0(u|_{\partial S}).$$

For the Dirichlet problems, these equalities produce the "substitute" integral equations

$$(W_0^* - \tfrac{1}{2}I)\varphi = N_0\mathcal{P}, \tag{$\mathcal{D}_\text{S}^+$}$$

$$(W_0^* + \tfrac{1}{2}I)\varphi = N_0\mathcal{R}. \tag{$\mathcal{D}_\text{S}^-$}$$

We bear in mind that in either case φ must still satisfy (2.111); that is,

$$p\varphi = 0. \tag{2.138}$$

It is obvious that $(\mathcal{D}_S^+)$, $(\mathcal{N}_D^-)$ and $(\mathcal{D}_S^-)$, $(\mathcal{N}_D^+)$ are mutually adjoint.

2.52. Theorem. (i) *The pair of equations $(\mathcal{D}_S^+)$ and (2.138) has a unique solution $\varphi \in C^{0,\alpha}(\partial S)$ for any $\mathcal{P} \in C^{1,\alpha}(\partial S)$, $\alpha \in (0,1)$.*
Then (D^+) has the (unique) solution

$$u = \mathcal{V}^+\varphi - \mathcal{W}^+\mathcal{P}. \tag{2.139}$$

(ii) *The pair of equations $(\mathcal{D}_S^-)$ and (2.138) has a unique solution $\varphi \in C^{0,\alpha}(\partial S)$ for any $\mathcal{R} \in C^{1,\alpha}(\partial S)$, $\alpha \in (0,1)$.*
Then (D^-) has the (unique) solution (in $\mathcal{A}^*$)

$$u = -\mathcal{V}^-\varphi + \mathcal{W}^-\mathcal{R} + F(q\mathcal{R}). \tag{2.140}$$

(iii) *$(\mathcal{N}_D^+)$ is solvable in $C^{1,\alpha}(\partial S)$ for any $\mathcal{Q} \in C^{0,\alpha}(\partial S)$ such that $p\mathcal{Q} = 0$, and its solution is unique up to an arbitrary rigid displacement Fa.*
Then (N^+) has the family of solutions

$$u = \mathcal{V}^+\mathcal{Q} - \mathcal{W}^+\psi + Fc, \tag{2.141}$$

where ψ is any solution of $(\mathcal{N}_D^+)$ and $c \in M_{3\times 1}$ is constant and arbitrary.
(iv) *$(\mathcal{N}_D^-)$ has a unique solution $\psi \in C^{1,\alpha}(\partial S)$ for any $\mathcal{S} \in C^{0,\alpha}(\partial S)$.*
Then, if $p\mathcal{S} = 0$, (N^-) has the (unique) solution (in $\mathcal{A}$)

$$u = -\mathcal{V}^-\mathcal{S} + \mathcal{W}^-\psi. \tag{2.142}$$

Proof. (i), (iv) The integral operators of $(\mathcal{D}_S^+)$ and $(\mathcal{N}_D^-)$ coincide with those of $(\mathcal{N}_C^-)$ and $(\mathcal{D}_C^+)$, respectively, so the unique solvability of $(\mathcal{D}_S^+)$ and $(\mathcal{N}_D^-)$ in $C^{0,\alpha}(\partial S)$ follows from the proof of Theorem 2.33(i),(iv). At the same time, operating with p in $(\mathcal{D}_S^+)$, we see that, by Lemma 2.32(ii),(iv), $p\varphi = 0$; in other words, the unique solution of $(\mathcal{D}_S^+)$ also satisfies (2.138), as required.

Now we need to check that (2.139) and (2.142), in which φ and ψ are the unique solutions of $(\mathcal{D}_S^+)$ and $(\mathcal{N}_D^-)$, satisfy (D^+) and (N^-), respectively, in accordance with Definition 2.4.

Since $\varphi \in C^{0,\alpha}(\partial S)$ and $\mathcal{P} \in C^{1,\alpha}(\partial S)$, the mapping properties of $\mathcal{V}^+$ and $\mathcal{W}^+$ (see Theorem 2.16) indicate that $\mathcal{V}^+\varphi, \mathcal{W}^+\mathcal{P} \in C^{1,\alpha}(\bar{S}^+)$, so u given by (2.139) belongs to $C^{1,\alpha}(\bar{S}^+)$. Also, $Au = 0$ (in S^+). All that remains to verify is that u satisfies the boundary condition of (D$^+$). Restricting (2.139) to ∂S, we have

$$u|_{\partial S} = V_0\varphi - (W_0 - \tfrac{1}{2}I)\mathcal{P}, \tag{2.143}$$

and we want to show that the right-hand side above is equal to $\mathcal{P}$. To this end, we consider the difference

$$\mathcal{G} = \left[V_0\varphi - (W_0 - \tfrac{1}{2}I)\mathcal{P}\right] - \mathcal{P} = V_0\varphi - (W_0 + \tfrac{1}{2}I)\mathcal{P}. \tag{2.144}$$

Applying N_0 and taking $(\mathcal{D}_S^+)$ and the compositions $(2.56)_2$ and $(2.57)_1$ into account, we find that

$$\begin{aligned}N_0\mathcal{G} &= (N_0V_0)\varphi - \bigl(N_0(W_0 + \tfrac{1}{2}I)\bigr)\mathcal{P} \\ &= (W_0^{*2} - \tfrac{1}{4}I)\varphi - (W_0^* + \tfrac{1}{2}I)(N_0\mathcal{P}) \\ &= (W_0^* + \tfrac{1}{2}I)\bigl[(W_0^* - \tfrac{1}{2}I)\varphi - N_0\mathcal{P}\bigr] = 0,\end{aligned}$$

so, by Theorem 2.24(ii),

$$\mathcal{G} = Fk \tag{2.145}$$

for some constant $k \in \mathcal{M}_{3\times 1}$. Operating with q in (2.144) and (2.145) and equating the results, taking into account (2.138) (which is satisfied by the solution φ of $(\mathcal{D}_S^+)$) and noting that $qF = (p\Phi)^{\mathrm{T}} = E_3$, we arrive at

$$k = q\mathcal{G} = q(V_0\varphi) - q(W_0\mathcal{P}) - \tfrac{1}{2}q\mathcal{P} = \mathcal{C}(p\varphi) + \tfrac{1}{2}q\mathcal{P} - \tfrac{1}{2}q\mathcal{P} = 0.$$

Now (2.145) yields $\mathcal{G} = 0$, which, combined with (2.143), shows that $u|_{\partial S} = \mathcal{P}$, as desired.

In the case of $(\mathcal{N}_D^-)$, since $\psi, \mathcal{S} \in C^{0,\alpha}(\partial S)$, it follows that, in fact,

$$\psi = 2(W_0\psi - V_0\mathcal{S}) \in C^{1,\alpha}(\partial S),$$

so $\mathcal{V}^-\mathcal{S}, \mathcal{W}^-\psi \in C^{1,\alpha}(\bar{S}^-)$; this means that u given by (2.142) belongs to $C^{1,\alpha}(\bar{S}^-)$. Also, $Au = 0$ (in S^-) and, if $p\mathcal{S} = 0$, then $u \in \mathcal{A}$ (by Theorem 2.14). To check that u satisfies the boundary condition of (N$^-$), we apply T in (2.142) and find that

$$Tu = -(W_0^* - \tfrac{1}{2}I)\mathcal{S} + N_0\psi. \tag{2.146}$$

Consider the function

$$\mathcal{H} = \bigl[-(W_0^* - \tfrac{1}{2}I)\mathcal{S} + N_0\psi\bigr] - \mathcal{S} = N_0\psi - (W_0^* + \tfrac{1}{2}I)\mathcal{S}. \tag{2.147}$$

Then
$$V_0\mathcal{H} = (V_0 N_0)\psi - \left(V_0(W_0^* + \tfrac{1}{2}I)\right)\mathcal{S}$$
$$= (W_0^2 - \tfrac{1}{4}I)\psi - (W_0 + \tfrac{1}{2}I)(V_0\mathcal{S})$$
$$= (W_0 + \tfrac{1}{2}I)\left[(W_0 - \tfrac{1}{2}I)\psi - V_0\mathcal{S}\right] = 0.$$

By Corollary 2.30,
$$\mathcal{H} = \Phi k, \tag{2.148}$$

where k is some constant (3×1)-vector such that $\mathcal{C}k = 0$. Applying p to (2.147) and (2.148) and using Lemma 2.32(ii),(iv), we see that

$$k = (p\Phi)k = p(\Phi k) = p\mathcal{H} = \tfrac{1}{2}p\mathcal{S} - \tfrac{1}{2}p\mathcal{S} = 0.$$

Consequently, $\mathcal{H} = 0$, which, in view of (2.147) and (2.146), shows that $Tu = \mathcal{S}$; in other words, u given by (2.142) is indeed the (unique) solution of (N^-).

(iii) The null space of the adjoint $W_0^* + \tfrac{1}{2}I$ of the operator on the left-hand side in $(\mathcal{N}_D^+)$ is spanned by $\{\Phi^{(i)}\}$. Also, by Lemma 2.32(i),

$$\int_{\partial S} \Phi^T (V_0 \mathcal{Q})\, ds = q(V_0 \mathcal{Q}) = \mathcal{C}(p\mathcal{Q}) = 0,$$

which is equivalent to
$$(\Phi^{(i)}, V_0\mathcal{Q}) = 0, \quad i = 1, 2, 3;$$

therefore, by the Fredholm Alternative, $(\mathcal{N}_D^+)$ is solvable in $C^{0,\alpha}(\partial S)$. Since the null space of $W_0 + \tfrac{1}{2}I$ is spanned by $\{F^{(i)}\}$, it follows that the solutions of $(\mathcal{N}_D^+)$ are

$$\psi = \bar{\psi} + Fc,$$

where $\bar{\psi}$ is any (fixed) solution.

As in the case of $(\mathcal{N}_D^-)$, here $\psi \in C^{1,\alpha}(\partial S)$, so u given by (2.141) belongs to $C^{1,\alpha}(\bar{S}^+)$. Since $Au = 0$ (in S^+), we only need to verify that u satisfies the boundary condition of (N^+). First, (2.141) yields

$$Tu = (W_0^* + \tfrac{1}{2}I)\mathcal{Q} - N_0\psi.$$

Next, the function
$$\mathcal{X} = \left[(W_0^* + \tfrac{1}{2}I)\mathcal{Q} - N_0\psi\right] - \mathcal{Q} = (W_0^* - \tfrac{1}{2}I)\mathcal{Q} - N_0\psi$$

satisfies

$$V_0 \mathcal{X} = \bigl(V_0(W_0^* - \tfrac{1}{2}I)\bigr)\mathcal{Q} - (V_0 N_0)\psi$$
$$= (W_0 - \tfrac{1}{2}I)(V_0 \mathcal{Q}) - (W_0^2 - \tfrac{1}{4}I)\psi$$
$$= (W_0 - \tfrac{1}{2}I)\bigl[V_0 \mathcal{Q} - (W_0 + \tfrac{1}{2}I)\psi\bigr] = 0,$$

and, since $p\mathcal{Q} = 0$,

$$p\mathcal{X} = p(W_0^* \mathcal{Q}) - \tfrac{1}{2}p\mathcal{Q} - p(N_0 \psi) = -p\mathcal{Q} = 0.$$

Combining these equalities, just as in (iv) we conclude that $\mathcal{X} = 0$, which means that $Tu = \mathcal{Q}$; that is, u given by (2.141) is a solution of (N^+) for any constant (3×1)-vector c.

(ii) Lemma 2.32(iv) implies that

$$\int_{\partial S} F^{\mathrm{T}}(N_0 \mathcal{R})\, ds = p(N_0 \mathcal{R}) = 0,$$

which is equivalent to

$$(F^{(i)}, N_0 \mathcal{R}) = 0, \quad i = 1, 2, 3;$$

so, by the Fredholm Alternative, $(\mathcal{D}_S^-)$ is solvable in $C^{0,\alpha}(\partial S)$ and its solutions are

$$\varphi = \bar{\varphi} + \Phi a, \tag{2.149}$$

where $\bar{\varphi}$ is any (fixed) solution and $a \in \mathcal{M}_{3 \times 1}$ is constant and arbitrary.

However, we need φ to satisfy (2.138); thus, operating with p in (2.149), we find that $a = -p\bar{\varphi}$, which reduces (2.149) to the solution

$$\varphi = \bar{\varphi} - \Phi(p\bar{\varphi}). \tag{2.150}$$

The difference $\tilde{\varphi}$ of two solutions of $(\mathcal{D}_S^-)$ and (2.138) satisfies

$$(W_0^* + \tfrac{1}{2}I)\tilde{\varphi} = 0, \quad p\tilde{\varphi} = 0.$$

The first equation tells us that, by Theorem 2.22(ii), $\tilde{\varphi} = \Phi a$ with some constant $a \in \mathcal{M}_{3 \times 1}$. But the second equation shows that $a = 0$, so the pair of equations $(\mathcal{D}_S^-)$ and (2.138) has a unique solution (2.150), irrespective of the solution $\bar{\varphi}$ of $(\mathcal{D}_S^-)$ chosen in its construction.

It remains to verify that u defined by (2.140) is the (unique) solution of (D^-). Since $\varphi \in C^{0,\alpha}(\partial S)$ and $\mathcal{R} \in C^{1,\alpha}(\partial S)$, it follows that $u \in C^{1,\alpha}(\bar{S}^-)$. In addition,

$Au = 0$ (in S^-) and (2.138) implies that $u \in \mathcal{A}^*$. Finally, the restriction of (2.140) to ∂S is written as

$$u|_{\partial S} = -V_0\varphi + (W_0 + \tfrac{1}{2}I)\mathcal{R} + F(q\mathcal{R}).$$

Setting

$$\mathcal{Y} = \bigl[-V_0\varphi + (W_0 + \tfrac{1}{2}I)\mathcal{R} + F(q\mathcal{R})\bigr] - \mathcal{R} = -V_0\varphi + (W_0 - \tfrac{1}{2}I)\mathcal{R} + F(q\mathcal{R}),$$

we find that

$$\begin{aligned} N_0\mathcal{Y} &= -(N_0V_0)\varphi + \bigl(N_0(W_0 - \tfrac{1}{2}I)\bigr)\mathcal{R} \\ &= -(W_0^{*2} - \tfrac{1}{4}I)\varphi + (W_0^* - \tfrac{1}{2}I)(N_0\mathcal{R}) \\ &= -(W_0^* - \tfrac{1}{2}I)\bigl[(W_0^* + \tfrac{1}{2}I)\varphi - N_0\mathcal{R}\bigr] = 0 \end{aligned}$$

and

$$q\mathcal{Y} = -q(V_0\varphi) + q(W_0\mathcal{R}) - \tfrac{1}{2}q\mathcal{R} + (qF)(q\mathcal{R}) = -\mathcal{C}(p\varphi) - \tfrac{1}{2}q\mathcal{R} - \tfrac{1}{2}q\mathcal{R} + q\mathcal{R} = 0.$$

Consequently, as in (i), we deduce that $u|_{\partial S} = \mathcal{R}$, which completes the proof.

2.53. Remark. The operators of $(\mathcal{D}_C^+)$, $(\mathcal{D}_C^-)$ are adjoint to those of $(\mathcal{D}_S^+)$, $(\mathcal{D}_S^-)$, and the operators of $(\mathcal{N}_C^+)$, $(\mathcal{N}_C^-)$ are adjoint to those of $(\mathcal{N}_D^+)$, $(\mathcal{N}_D^-)$.

2.54. Remark. We can also devise a substitute method for $(R^\pm)$, where we can choose Tu as the unknown function in the corresponding boundary integral equations.

Thus, the Somigliana formula $(2.52)_1$ restricted to ∂S yields

$$V_0(Tu) - (W_0 + \tfrac{1}{2}I)(u|_{\partial S}) = 0.$$

Writing

$$\theta = (Tu)/\sigma, \qquad (2.151)$$

from the boundary condition of (R^+) we have

$$u|_{\partial S} = -\theta + \mathcal{K}/\sigma,$$

and the above equation becomes

$$V_0(\sigma\theta) + (W_0 + \tfrac{1}{2}I)\theta = (W_0 + \tfrac{1}{2}I)(\mathcal{K}/\sigma). \qquad (\mathcal{R}_S^+)$$

Similarly, restricting the representation (2.110) to ∂S, we obtain

$$V_0(Tu) - (W_0 - \tfrac{1}{2}I)(u|_{\partial S}) - Fc = 0,$$

which, under the substitution

$$u|_{\partial S} = \theta - \mathcal{L}/\sigma,$$

where θ is given by (2.151), reduces to

$$V_0(\sigma\theta) - (W_0 - \tfrac{1}{2}I)\theta - Fc = -(W_0 - \tfrac{1}{2}I)(\mathcal{L}/\sigma). \qquad (\mathcal{R}_S^-)$$

As in the direct method, since $\sigma\theta = Tu$, it follows that in both the interior and exterior problems θ must satisfy the additional restriction

$$p(\sigma\theta) = 0. \qquad (2.152)$$

We remark that the operators on the left-hand side in $(\mathcal{R}_S^+)$ and $(\mathcal{R}_S^-)$ are the same as those in $(\mathcal{R}_D^+)$ and $(\mathcal{R}_D^-)$, respectively, so the handling of the substitute equations is similar to that of the equations in the direct method.

Operating with q in $(\mathcal{R}_S^+)$ and using Lemma 2.32(i),(iii), we arrive at

$$\mathcal{C}p(\sigma\theta) = 0,$$

so, if $\det \mathcal{C} \neq 0$, then (2.152) is satisfied by the (unique) solution of $(\mathcal{R}_S^+)$.

Doing the same in $(\mathcal{R}_S^-)$, we obtain

$$\mathcal{C}p(\sigma\theta) + q\theta = c + q(\mathcal{L}/\sigma),$$

or

$$c = \mathcal{C}p(\sigma\theta) + q(\theta - \mathcal{L}/\sigma). \qquad (2.153)$$

Following the procedure in §2.11, we infer that $(\mathcal{R}_S^-)$ has a unique solution for

$$c = q(\theta - \mathcal{L}/\sigma).$$

Then (2.153) shows that when $\det \mathcal{C} \neq 0$, this solution also satisfies (2.152).

Just as in the proof of Theorem 2.50, it can now be verified that the functions

$$u = \mathcal{V}^+(\sigma\theta) - \mathcal{W}^+(-\theta + \mathcal{K}/\sigma),$$
$$u = -\mathcal{V}^-(\sigma\theta) + \mathcal{W}^+(\theta - \mathcal{L}/\sigma) + Fq(\theta - \mathcal{L}/\sigma),$$

suggested by the Somigliana representation formulae $(2.52)_1$ and (2.110), are the (unique) solutions of (R^+) and (R^-), respectively.

Chapter 3
Bending of Elastic Plates

3.1. Notation and prerequisites

In this chapter we adopt all the symbols and conventions introduced in Chapter 2.

Let S be a domain in $\mathbb{R}^2$ bounded by a simple closed C^2-curve ∂S, and let $h_0 = \text{const}$ be such that $0 < h_0 \ll \text{diam}\, S$. By a *thin plate* we understand an elastic body that occupies the region $\Omega = \bar{S} \times [-h_0/2, h_0/2]$; here h_0 is called the *thickness* of the plate.

For functions s defined in Ω we consider the averaging operators $\mathcal{I}_0$, $\mathcal{I}_1$ and $\mathcal{J}_0$, $\mathcal{J}_1$ given by

$$(\mathcal{I}_0 s)(x_\gamma) = h_0^{-1} \big[s(x_i)\big]_{x_3=-h_0/2}^{x_3=h_0/2},$$

$$(\mathcal{I}_1 s)(x_\gamma) = h_0^{-1} \big[x_3 s(x_i)\big]_{x_3=-h_0/2}^{x_3=h_0/2},$$

$$(\mathcal{J}_0 s)(x_\gamma) = h_0^{-1} \int_{-h_0/2}^{h_0/2} s(x_i)\, dx_3, \qquad (3.1)$$

$$(\mathcal{J}_1 s)(x_\gamma) = h_0^{-1} \int_{-h_0/2}^{h_0/2} x_3 s(x_i)\, dx_3$$

and write

$$N_{\alpha\beta} = \mathcal{J}_1 t_{\alpha\beta}, \quad N_{3\alpha} = \mathcal{J}_0 t_{3\alpha},$$
$$g_\alpha = \mathcal{J}_1 f_\alpha + \mathcal{I}_1 t_{3\alpha}, \quad g_3 = \mathcal{J}_0 f_3 + \mathcal{I}_0 t_{33}. \qquad (3.2)$$

It can be shown that $N_{\alpha\alpha}$ (α not summed), $N_{12} = N_{21}$ and $N_{3\alpha}$ are the bending and twisting moments (with respect to the middle plane $x_3 = 0$) and transverse shear forces acting on a vertical cross-section element of the plate perpendicular to the x_α-axis, all averaged over the thickness, and that $\mathcal{J}_0 f_3$, $\mathcal{J}_1 f_\alpha$ and $\mathcal{I}_0 t_{33}$, $\mathcal{I}_1 t_{3\alpha}$ are the averaged body force and moments and the averaged force and moments acting on the faces $x_3 = \pm h_0/2$, respectively. Since there is no danger of ambiguity, in what follows we omit the word 'averaged' when we refer to these forces and moments.

Writing
$$N_\alpha = N_{\alpha\beta} \nu_\beta, \quad N_3 = N_{3\beta} \nu_\beta, \qquad (3.3)$$

where $\nu = (\nu_1, \nu_2)^{\mathrm{T}}$ is the unit normal on ∂S pointing towards the far field, we can also show that the components of the moment with respect to the middle plane and those of the transverse shear force on the boundary are $\varepsilon_{\beta\alpha} N_\beta$ and N_3. If, on the other hand, the moment is computed with respect to the origin, then its components are $\varepsilon_{\beta\alpha}(N_\beta - x_\beta N_3)$. It is obvious that prescribing the N_i on ∂S is equivalent to prescribing the $\varepsilon_{\beta\alpha}(N_\beta - x_\beta N_3)$ and N_3.

In the model of bending of plates with transverse shear deformation it is assumed that the displacements are of the form

$$x_3 u_\alpha(x_1, x_2), \quad u_3(x_1, x_2). \tag{3.4}$$

It should be emphasized that here

$$u = (u_1, u_2, u_3)^\mathrm{T}$$

is not the displacement vector itself, but a vector that characterizes the displacements uniquely in accordance with (3.4). However, for simplicity we will refer to the u_i as the displacements.

Applying $\mathcal{J}_1$ to $(2.1)_\alpha$ and $(2.2)_\alpha$, and $\mathcal{J}_0$ to $(2.1)_3$, $(2.2)_3$ and (2.4), from (3.1) and (3.2) we now obtain the equilibrium equations, constitutive relations and internal energy density, respectively, in the form

$$\begin{aligned} N_{\alpha\beta,\beta} - N_{3\alpha} &= g_\alpha, \\ N_{3\beta,\beta} &= g_3, \end{aligned} \tag{3.5}$$

$$\begin{aligned} N_{\alpha\beta} &= h^2\bigl[\lambda u_{\gamma,\gamma}\delta_{\alpha\beta} + \mu(u_{\alpha,\beta} + u_{\beta,\alpha})\bigr], \\ N_{3\alpha} &= \mu(u_\alpha + u_{3,\alpha}), \end{aligned} \tag{3.6}$$

$$E = \tfrac{1}{4} N_{\alpha\beta}(u_{\alpha,\beta} + u_{\beta,\alpha}) + \tfrac{1}{2} N_{3\alpha}(u_\alpha + u_{3,\alpha}), \tag{3.7}$$

where

$$h^2 = h_0^2/12.$$

We can also write the equilibrium equations in terms of the displacements. Thus, substituting (3.6) in (3.5), we arrive at

$$\begin{aligned} h^2(\lambda+\mu)u_{\beta,\beta\alpha} + h^2\mu u_{\alpha,\beta\beta} - \mu(u_\alpha + u_{3,\alpha}) + g_\alpha &= 0, \\ \mu u_{\beta,\beta} + \mu u_{3,\beta\beta} + g_3 &= 0, \end{aligned} \tag{3.8}$$

or

$$\begin{aligned} h^2(\lambda+\mu)\operatorname{grad}\operatorname{div}\bar{u} + h^2\mu\Delta\bar{u} - \mu\bar{u} - \mu\operatorname{grad} u_3 + \bar{g} &= 0, \\ \mu \operatorname{div}\bar{u} + \mu\Delta u_3 + g_3 &= 0, \end{aligned}$$

where

$$\bar{u} = (u_1, u_2)^\mathrm{T}, \quad \bar{g} = (g_1, g_2)^\mathrm{T}.$$

For the purpose of our analysis we assume that the body forces and moments and the forces and moments acting on the faces of the plate are negligible (in other words, $g_i = 0$) and write (3.8) as

$$A(\partial_1, \partial_2)u = 0, \tag{3.9}$$

where $A(\partial_1, \partial_2)$ is the matrix differential operator

$$\begin{pmatrix} h^2\mu\Delta + h^2(\lambda+\mu)\partial_1^2 - \mu & h^2(\lambda+\mu)\partial_1\partial_2 & -\mu\partial_1 \\ h^2(\lambda+\mu)\partial_1\partial_2 & h^2\mu\Delta + h^2(\lambda+\mu)\partial_2^2 - \mu & -\mu\partial_2 \\ \mu\partial_1 & \mu\partial_2 & \mu\Delta \end{pmatrix}. \tag{3.10}$$

As in all two-dimensional problems concerning homogeneous and isotropic materials, in what follows we assume that

$$\lambda + \mu > 0, \quad \mu > 0. \tag{3.11}$$

3.1. Theorem. *System (3.9) is elliptic.*

Proof. Considering the matrix

$$A_0(\xi_1, \xi_2) = \begin{pmatrix} h^2\mu\xi_\alpha\xi_\alpha + h^2(\lambda+\mu)\xi_1^2 & h^2(\lambda+\mu)\xi_1\xi_2 & 0 \\ h^2(\lambda+\mu)\xi_1\xi_2 & h^2\mu\xi_\alpha\xi_\alpha + h^2(\lambda+\mu)\xi_2^2 & 0 \\ 0 & 0 & \mu\xi_\alpha\xi_\alpha \end{pmatrix}$$

corresponding to the second order derivatives in (3.9), we see that

$$\det A_0(\xi_1, \xi_2) = h^4\mu^2(\lambda+2\mu)(\xi_1^2 + \xi_2^2)^3 > 0 \quad \text{for all } (\xi_1, \xi_2) \neq 0;$$

therefore, $A_0(\xi_1, \xi_2)$ is invertible for all $(\xi_1, \xi_2) \neq 0$, which means that (3.9) is an elliptic system.

The moments and transverse shear force on the boundary (see (3.3)) can be written in the form

$$N_i = N_{i\beta}\nu_\beta = T_{ij}u_j,$$

where, with (3.6) replaced in (3.3), the boundary matrix differential operator $T(\partial_1, \partial_2)$ is

$$\begin{pmatrix} h^2(\lambda+2\mu)\nu_1\partial_1 + h^2\mu\nu_2\partial_2 & h^2\mu\nu_2\partial_1 + h^2\lambda\nu_1\partial_2 & 0 \\ h^2\lambda\nu_2\partial_1 + h^2\mu\nu_1\partial_2 & h^2\mu\nu_1\partial_1 + h^2(\lambda+2\mu)\nu_2\partial_2 & 0 \\ \mu\nu_1 & \mu\nu_2 & \mu\nu_\alpha\partial_\alpha \end{pmatrix}. \tag{3.12}$$

Finally, from (3.7) and (3.6) we obtain the expression of the internal energy density per unit area of the middle plane:

$$E = E(u, u) = \tfrac{1}{2}\{h^2[\lambda u_{\alpha,\alpha} u_{\beta,\beta} + \mu u_{\alpha,\beta}(u_{\alpha,\beta} + u_{\beta,\alpha})] \\ + \mu(u_\alpha + u_{3,\alpha})(u_\alpha + u_{3,\alpha})\}. \qquad (3.13)$$

Comparing the general form of a three-dimensional displacement

$$(a_1 + b_3 x_2 - b_2 x_3,\ a_2 + b_1 x_3 - b_3 x_1,\ a_3 + b_2 x_1 - b_1 x_2)^{\mathrm{T}}$$

with (3.4), we deduce that we must have

$$a_1 = a_2 = b_3 = 0;$$

that is, an arbitrary admissible rigid displacement for this model is of the form

$$u(x) = (c_1,\ c_2,\ c_0 - c_1 x_1 - c_2 x_2)^{\mathrm{T}}, \qquad (3.14)$$

where c_1, c_2 and c_0 are constants. We denote by $\mathcal{F}$ the vector space of all such displacements and remark that the columns $F^{(i)}$ of the matrix

$$F = \begin{pmatrix} 1 & 0 & 0 \\ 0 & 1 & 0 \\ -x_1 & -x_2 & 1 \end{pmatrix} \qquad (3.15)$$

are linearly independent and span $\mathcal{F}$; hence, they form a basis for $\mathcal{F}$. It is also easily verified that

$$(AF)(x) = 0, \quad x \in \mathbb{R}^2,$$
$$(TF)(x) = 0, \quad x \in \partial S.$$

With the notation

$$k_1 = c_1, \quad k_2 = c_2, \quad k_3 = c_0,$$

we can write the general admissible rigid displacement (3.14) in the form Fk, where $k \in \mathcal{M}_{3\times 1}$ is constant and arbitrary.

3.2. Theorem. $E(u, u)$ is a positive quadratic form, and $E(u, u) = 0$ if and only if $u \in \mathcal{F}$.

Proof. We write (3.13) as

$$E(u, u) = \tfrac{1}{2}\{h^2[E_0(u, u) + \mu(u_{1,2} + u_{2,1})^2] + \mu[(u_1 + u_{3,1})^2 + (u_2 + u_{3,2})^2]\},$$

where, as seen in the proof of Theorem 2.2,

$$E_0(u, u) = (\lambda + 2\mu)u_{1,1}^2 + 2\lambda u_{1,1} u_{2,2} + (\lambda + 2\mu)u_{2,2}^2$$

is a positive quadratic form; consequently, so is $E(u, u)$.

Clearly, $E(u, u) = 0$ for u of the form (3.14). Conversely, $E(u, u) = 0$ implies that

$$\begin{aligned} u_{1,1} &= u_{2,2} = 0, \\ u_{1,2} + u_{2,1} &= 0, \\ u_{3,\alpha} + u_\alpha &= 0, \end{aligned} \tag{3.16}$$

so

$$\begin{aligned} u_1 &= s_1(x_2), \\ u_2 &= s_2(x_1), \end{aligned} \tag{3.17}$$

which, when used in $(3.16)_2$, yield

$$\begin{aligned} s_1(x_2) &= ax_2 + c_1, \\ s_2(x_1) &= -ax_1 + c_2, \end{aligned}$$

where a, c_1 and c_2 are arbitrary constants. Since $u_{3,12} = u_{3,21}$, from the last two equations we find that $a = 0$; therefore, by (3.17),

$$u_\alpha = c_\alpha,$$

and from $(3.16)_3$ it follows that

$$u_{3,\alpha} = -c_\alpha,$$

which finally leads us to

$$u_3 = c_0 - c_\alpha x_\alpha,$$

where c_0 is an arbitrary constant.

3.3. Lemma. *If $u \in C^2(S^+) \cap C^1(\bar{S}^+)$, then*

$$\int_{S^+} F^{\mathrm{T}}(Au)\, da = \int_{\partial S} F^{\mathrm{T}}(Tu)\, ds. \tag{3.18}$$

Proof. Given the equivalence of (3.9) and (3.5) (with $g_i = 0$), we use the divergence theorem to obtain the relations

$$\int_{S^+} (A_{\alpha i} - x_\alpha A_{3i}) u_i \, da = \int_{S^+} (N_{\alpha\beta,\beta} - N_{3\alpha} - x_\alpha N_{3\beta,\beta}) \, da$$

$$= \int_{S^+} (N_{\alpha\beta} - x_\alpha N_{3\beta})_{,\beta} \, da = \int_{\partial S} (N_{\alpha\beta} - x_\alpha N_{3\beta}) \nu_\beta \, ds = \int_{\partial S} (T_{\alpha i} - x_\alpha T_{3i}) u_i \, ds,$$

$$\int_{S^+} A_{3i} u_i \, da = \int_{S^+} N_{3\beta,\beta} \, da = \int_{\partial S} N_{3\beta} \nu_\beta \, ds = \int_{\partial S} T_{3i} u_i \, ds,$$

which, in view of (3.15), are equivalent to the required equality.

3.2. The fundamental boundary value problems

We denote by $\mathcal{A}$ the space of all functions $u \in \mathcal{M}_{3\times 1}$ defined in S^- which, in polar coordinates r, θ with the pole at the origin, have, as $r = |x| \to \infty$, an asymptotic expansion of the form

$$u_1(r,\theta) = r^{-1}\big[m_0 \sin\theta + 2m_1 \cos\theta - m_0 \sin 3\theta + (m_2 - m_1)\cos 3\theta\big]$$
$$+ r^{-2}\big[(2m_3 + m_4)\sin 2\theta + m_5 \cos 2\theta - 2m_3 \sin 4\theta + 2m_6 \cos 4\theta\big]$$
$$+ r^{-3}\big[2m_7 \sin 3\theta + 2m_8 \cos 3\theta + 3(m_9 - m_7)\sin 5\theta$$
$$+ 3(m_{10} - m_8)\cos 5\theta\big] + \tilde{u}_1(r,\theta),$$

$$u_2(r,\theta) = r^{-1}\big[2m_2 \sin\theta + m_0 \cos\theta + (m_2 - m_1)\sin 3\theta + m_0 \cos 3\theta\big]$$
$$+ r^{-2}\big[(2m_6 + m_5)\sin 2\theta - m_4 \cos 2\theta + 2m_6 \sin 4\theta + 2m_3 \cos 4\theta\big]$$
$$+ r^{-3}\big[2m_{10}\sin 3\theta - 2m_9 \cos 3\theta + 3(m_{10} - m_8)\sin 5\theta \qquad (3.19)$$
$$+ 3(m_7 - m_9)\cos 5\theta\big] + \tilde{u}_2(r,\theta),$$

$$u_3(r,\theta) = -(m_1 + m_2)\ln r - \big[m_1 + m_2 + m_0 \sin 2\theta + (m_1 - m_2)\cos 2\theta\big]$$
$$+ r^{-1}\big[(m_3 + m_4)\sin\theta + (m_5 + m_6)\cos\theta - m_3 \sin 3\theta + m_6 \cos 3\theta\big]$$
$$+ r^{-2}\big[m_{11}\sin 2\theta + m_{12}\cos 2\theta + (m_9 - m_7)\sin 4\theta$$
$$+ (m_{10} - m_8)\cos 4\theta\big] + \tilde{u}_3(r,\theta),$$

where $m_1, \ldots, m_{12}$ are arbitrary constants and

$$\tilde{u}_\alpha = O(r^{-4}), \quad \tilde{u}_3 = O(r^{-3}),$$
$$\partial_r \tilde{u}_\alpha = O(r^{-5}), \quad \partial_r \tilde{u}_3 = O(r^{-4}), \qquad (3.20)$$
$$\partial_\theta \tilde{u}_\alpha = O(r^{-4}), \quad \partial_\theta \tilde{u}_3 = O(r^{-3}),$$

uniformly with respect to θ.

We also define
$$\mathcal{A}^* = \mathcal{A} \oplus \mathcal{F},$$
that is, the set of all functions of the form $u = u^{\mathcal{A}} + Fk$, where $u^{\mathcal{A}} \in \mathcal{A}$ and $k \in \mathcal{M}_{3\times 1}$ is constant and arbitrary.

As is clear from (3.13), $\mathcal{A}$ and $\mathcal{A}^*$ are classes of finite energy functions.

3.4. Definition. Let $\mathcal{P}$, $\mathcal{Q}$, $\mathcal{R}$, $\mathcal{S}$, $\mathcal{K}$, $\mathcal{L} \in C(\partial S)$ be given (3×1)-matrix functions, and let $\sigma \in C(\partial S)$ be a given positive definite (3×3)-matrix function. We consider the following interior and exterior Dirichlet, Neumann and Robin problems.

(D$^+$) Find $u \in C^2(S^+) \cap C^1(\bar{S}^+)$ such that
$$(Au)(x) = 0, \quad x \in S^+,$$
$$u(x) = \mathcal{P}(x), \quad x \in \partial S.$$

(N$^+$) Find $u \in C^2(S^+) \cap C^1(\bar{S}^+)$ such that
$$(Au)(x) = 0, \quad x \in S^+,$$
$$(Tu)(x) = \mathcal{Q}(x), \quad x \in \partial S.$$

(R$^+$) Find $u \in C^2(S^+) \cap C^1(\bar{S}^+)$ such that
$$(Au)(x) = 0, \quad x \in S^+,$$
$$(Tu + \sigma u)(x) = \mathcal{K}(x), \quad x \in \partial S.$$

(D$^-$) Find $u \in C^2(S^-) \cap C^1(\bar{S}^-) \cap \mathcal{A}^*$ such that
$$(Au)(x) = 0, \quad x \in S^-,$$
$$u(x) = \mathcal{R}(x), \quad x \in \partial S.$$

(N$^-$) Find $u \in C^2(S^-) \cap C^1(\bar{S}^-) \cap \mathcal{A}$ such that
$$(Au)(x) = 0, \quad x \in S^-,$$
$$(Tu)(x) = \mathcal{S}(x), \quad x \in \partial S.$$

(R$^-$) Find $u \in C^2(S^-) \cap C^1(\bar{S}^-) \cap \mathcal{A}^*$ such that
$$(Au)(x) = 0, \quad x \in S^-,$$
$$(Tu - \sigma u)(x) = \mathcal{L}(x), \quad x \in \partial S.$$

A function satisfying any one of the above sets of equations is called a *regular solution* of that boundary value problem, or just a *solution*.

3.5. Remark. The above boundary value problems for the non-homogeneous system (3.9) can be reduced to those in Definition 3.4 by means of a suitably constructed particular solution of the non-homogeneous system (see §3.7).

3.6. Theorem. *If* (N^+) *and* (N^-) *are solvable, then*

$$p\mathcal{Q} = 0, \quad p\mathcal{S} = 0,$$

respectively, where p is the operator defined on $C(\partial S)$ *by*

$$p\varphi = \int_{\partial S} F^\mathrm{T} \varphi \, ds.$$

Proof. The first equality is derived as in the case of Theorem 2.6, by means of (3.18). In the same way we establish the analogue of formula (2.25), where we need to show that

$$\int_{\partial K_R} F^\mathrm{T}(Tu) \, ds \to 0 \quad \text{as } R \to \infty. \tag{3.21}$$

Bearing in mind that $u \in \mathcal{A}$ and using (2.26), (3.12), (3.19) and (3.20), we see that

$$\begin{aligned}(Tu)_\alpha &= T_{\alpha\beta} u_\beta + T_{\alpha 3} u_3 = T_{\alpha\beta} u_\beta = O(R^{-2}),\\ (Tu)_3 &= T_{3\beta} u_\beta + T_{33} u_3 = O(R^{-1}),\end{aligned} \tag{3.22}$$

so

$$\begin{aligned}\left(F^\mathrm{T}(Tu)\right)_\alpha &= F_{\beta\alpha}(Tu)_\beta + F_{3\alpha}(Tu)_3 \\ &= \delta_{\beta\alpha}(Tu)_\beta - x_\alpha (Tu)_3 = (Tu)_\alpha - x_\alpha(Tu)_3 = O(1),\\ \left(F^\mathrm{T}(Tu)\right)_3 &= F_{\beta 3}(Tu)_\beta + F_{33}(Tu)_3 = (Tu)_3 = O(R^{-1}),\end{aligned}$$

which would seem to indicate that (3.21) does not hold. This is explained by the coarseness of (3.22). If we make use of the full detailed structure of the expansion (3.19) instead of considering just the order of magnitude of the terms, after a long but straightforward calculation we obtain the more refined estimate

$$(Tu)_3 = O(R^{-3}). \tag{3.23}$$

Consequently, we now find that

$$F^{\mathrm{T}}(Tu) = O(R^{-2}),$$

from which (3.21) follows immediately.

3.3. The Betti and Somigliana formulae

These equalities are similar in notation to those established in Chapter 2, but the symbols involved represent different quantities and the proofs require more accurate estimates.

3.7. Theorem. *If* $u \in C^2(S^+) \cap C^1(\bar{S}^+)$, *then*

$$\int_{S^+} u^{\mathrm{T}}(Au)\, da + 2\int_{S^+} E(u,u)\, da = \int_{\partial S} u^{\mathrm{T}}(Tu)\, ds. \tag{3.24}$$

Proof. Using the divergence theorem, (3.7) and the equivalence of (3.9) and (3.5) (with $g_i = 0$), we find that

$$\int_{S^+} u^{\mathrm{T}}(Au)\, da = \int_{S^+} u_i(Au)_i\, da = \int_{S^+} \left[u_\alpha(N_{\alpha\beta,\beta} - N_{3\alpha}) + u_3 N_{3\beta,\beta}\right] da$$

$$= \int_{S^+} \left[u_\alpha N_{\alpha\beta} + u_3 N_{3\beta})_{,\beta} - N_{\alpha\beta}u_{\alpha,\beta} - N_{3\alpha}(u_\alpha + u_{3,\alpha})\right] da$$

$$= \int_{\partial S} u_i N_{i\beta}\nu_\beta\, ds - \int_{S^+} \left[N_{\alpha\beta}u_{\alpha,\beta} + N_{3\alpha}(u_\alpha + u_{3,\alpha})\right] da$$

$$= \int_{\partial S} u^{\mathrm{T}}(Tu)\, ds - 2\int_{S^+} E(u,u)\, da,$$

which yields (3.24).

As in plane strain, the Betti formulae follow directly from Theorem 3.7.

3.8. Corollary. (i) *If* $u \in C^2(S^+) \cap C^1(\bar{S}^+)$ *satisfies* $(Au)(x) = 0$, $x \in S^+$, *then*

$$2\int_{S^+} E(u,u)\, da = \int_{\partial S} u^{\mathrm{T}}(Tu)\, ds.$$

(ii) *If* $u \in C^2(S^-) \cap C^1(\bar{S}^-) \cap \mathcal{A}^*$ *satisfies* $(Au)(x) = 0$, $x \in S^-$, *then*

$$2\int_{S^-} E(u,u)\, da = -\int_{\partial S} u^{\mathrm{T}}(Tu)\, ds.$$

Proof. (i) This is a direct consequence of (3.24).

(ii) As in the proof of Corollary 2.8(ii), we derive the equality

$$2\int_{S^-\cap K_R} E(u,u)\,da = \left(-\int_{\partial S} + \int_{\partial K_R}\right)u^{\mathrm{T}}(Tu)\,ds$$

and must show that

$$\int_{\partial K_R} u^{\mathrm{T}}(Tu)\,ds \to 0 \quad \text{as } R \to \infty. \tag{3.25}$$

Since $TF = 0$, we see that, by (3.14), (3.19), (3.22)$_1$ and (3.23), for $u \in \mathcal{A}^*$ we have

$$u^{\mathrm{T}}(Tu) = u_\alpha(Tu)_\alpha + u_3(Tu)_3 = O(1)O(R^{-2}) + O(R)O(R^{-3}) = O(R^{-2}),$$

which means that (3.25) holds and we obtain the desired formula.

3.9. Theorem. *If $u, v \in C^2(S^+) \cap C^1(\bar{S}^+)$, then*

$$\int_{S^+} \left[u^{\mathrm{T}}(Av) - v^{\mathrm{T}}(Au)\right] da = \int_{\partial S} \left[u^{\mathrm{T}}(Tv) - v^{\mathrm{T}}(Tu)\right] ds. \tag{3.26}$$

Proof. Denoting by $S_{i\beta}$ the moments and shear forces generated by the v_i, we can write

$$\int_{S^+} \left[u^{\mathrm{T}}(Av) - v^{\mathrm{T}}(Au)\right] da$$

$$= \int_{S^+} \left[u_\alpha(S_{\alpha\beta,\beta} - S_{3\alpha}) + u_3 S_{3\beta,\beta} - v_\alpha(N_{\alpha\beta,\beta} - N_{3\alpha}) - v_3 N_{3\beta,\beta}\right] da$$

$$= \int_{S^+} \left[(u_\alpha S_{\alpha\beta} + u_3 S_{3\beta})_{,\beta} - (v_\alpha N_{\alpha\beta} + v_3 N_{3\beta})_{,\beta}\right.$$

$$\left. - u_{\alpha,\beta} S_{\alpha\beta} + v_{\alpha,\beta} N_{\alpha\beta} - (u_\alpha + u_{3,\alpha})S_{3\alpha} + (v_\alpha + v_{3,\alpha})N_{3\alpha}\right] da$$

$$= \int_{\partial S} (u_i S_{i\beta} - v_i N_{i\beta})\nu_\beta\,ds - \int_{S^+} \mathcal{B}(u,v)\,da.$$

Using (3.6), we easily check that, in fact, $\mathcal{B}(u,v) = 0$, so (3.26) holds.

We can obtain a matrix of fundamental solutions for $-A$ by writing

$$D(x,y) = (\operatorname{adj} A)(\partial_x)\left[t(x,y)E_3\right] \tag{3.27}$$

in place of u in (3.9); here ∂_x signifies action with respect to x. Applying $A(\partial_x)$ to (3.27), we find that

$$A(\partial_x)D(x,y) = A(\partial_x)(\operatorname{adj} A)(\partial_x)\bigl[t(x,y)E_3\bigr] = (\det A)(\partial_x)\bigl[t(x,y)E_3\bigr], \qquad (3.28)$$

so $D(x,y)$ is a matrix of fundamental solutions for $-A$ if t satisfies

$$(\det A)(\partial_x)t(x,y) = -\delta(|x-y|); \qquad (3.29)$$

that is, t is a fundamental solution for the scalar operator $-\det A$.

Direct computation based on (3.10) leads to the expression

$$(\det A)(\partial_1, \partial_2) = h^4\mu^2(\lambda + 2\mu)\Delta\Delta(\Delta - h^{-2}),$$

so t must be a solution of the equation

$$\Delta\Delta(\Delta - h^{-2})(\partial_x)t(x,y) = -\bigl[h^4\mu^2(\lambda + 2\mu)\bigr]^{-1}\delta(|x-y|). \qquad (3.30)$$

We seek t as a linear combination of the fundamental solutions of Δ, $\Delta\Delta$ and $\Delta - h^{-2}$; in other words,

$$t(x,y) = b_1 \ln|x-y| + b_2|x-y|^2 \ln|x-y| + b_3 K_0(h^{-1}|x-y|), \qquad (3.31)$$

where K_0 is the modified Bessel function of the first kind and order zero. Recalling that

$$\Delta(\partial_x)(\ln|x-y|) = 2\pi\delta(|x-y|),$$
$$(\Delta\Delta)(\partial_x)(|x-y|^2 \ln|x-y|) = 8\pi\delta(|x-y|),$$
$$(\Delta - h^{-2})(\partial_x)K_0(h^{-1}|x-y|) = -2\pi\delta(|x-y|),$$

we apply $\Delta(\partial_x)$ to (3.31) and arrive at

$$\Delta(\partial_x)t(x,y) = 2\pi(b_1 - b_3)\delta(|x-y|) + 4b_2(1 + \ln|x-y|)$$
$$+ b_3 h^{-2} K_0(h^{-1}|x-y|), \qquad (3.32)$$

so asking that this expression should contain no δ-terms leads to

$$b_1 = b_3. \qquad (3.33)$$

We repeat the procedure in (3.32) and find that

$$(\Delta\Delta)(\partial_x)t(x,y) = 2\pi(4b_2 - b_3 h^{-2})\delta(|x-y|) + b_3 h^{-4} K_0(h^{-1}|x-y|); \quad (3.34)$$

therefore,

$$b_2 = \tfrac{1}{4} b_3 h^{-2}. \quad (3.35)$$

Finally, applying the operator $\Delta - h^{-2}$ to both sides in (3.34) and taking (3.30) into account, we deduce that

$$-2\pi b_3 h^{-4}\delta(|x-y|) = -\bigl[h^4\mu^2(\lambda + 2\mu)\bigr]^{-1}\delta(|x-y|);$$

hence,

$$b_3 = \bigl[2\pi\mu^2(\lambda + 2\mu)\bigr]^{-1}. \quad (3.36)$$

From (3.31), (3.35) and (3.36) we now obtain

$$t(x,y) = t(|x-y|) = a\bigl[(4h^2 + |x-y|^2)\ln|x-y| + 4h^2 K_0(h^{-1}|x-y|)\bigr], \quad (3.37)$$

where

$$a = \bigl[8\pi h^2\mu^2(\lambda + 2\mu)\bigr]^{-1}.$$

This means that a matrix of fundamental solutions for $-A$ is $D(x,y)$ given by (3.27), where t is as in (3.37) and the entries of adj A, computed directly from (3.10), are

$$\begin{aligned}
(\operatorname{adj} A)_{\alpha\beta}(\partial_1, \partial_2) &= h^2\mu(\lambda + 2\mu)\delta_{\alpha\beta}\Delta\Delta - h^2\mu(\lambda + \mu)\Delta\partial_\alpha\partial_\beta - \mu^2\partial_\alpha\partial_\beta, \\
(\operatorname{adj} A)_{33}(\partial_1, \partial_2) &= h^4\mu(\lambda + 2\mu)\Delta\Delta - h^2\mu(\lambda + 3\mu)\Delta + \mu^2, \\
(\operatorname{adj} A)_{\alpha 3}(\partial_1, \partial_2) &= -(\operatorname{adj} A)_{3\alpha}(\partial_1, \partial_2) = \mu^2\partial_\alpha(h^2\Delta - 1).
\end{aligned} \quad (3.38)$$

From (3.27), (3.37) and (3.38) it follows that

$$D(x,y) = D^{\mathrm{T}}(y,x). \quad (3.39)$$

Another important matrix of singular solutions is

$$P(x,y) = \bigl[T(\partial_y)D(y,x)\bigr]^{\mathrm{T}}. \quad (3.40)$$

As in Chapter 2, it is easily shown that

$$A(\partial_x)D(x,y) = 0, \quad A(\partial_x)P(x,y) = 0, \quad x \in \mathbb{R}^2, \ x \neq y;$$

that is, the columns $D^{(i)}$ and $P^{(i)}$ of D and P are solutions of (3.9) at all points $x \neq y$.

It is essential that we know the behaviour of both $D(x,y)$ and $P(x,y)$ near $x = y$. Since, as $\xi \to 0+$, we have [1]

$$K_0(\xi) = -(1 + \tfrac{1}{4}\xi^2 + \tfrac{1}{64}\xi^4 + \cdots)\ln\xi,$$

we use (3.31) to find that for x close to y

$$t(x,y) = -\left[128\pi h^4\mu^2(\lambda + 2\mu)\right]^{-1}|x-y|^4 \ln|x-y| + \tilde{t}(x,y),$$

where $\tilde{t} \in C^5(\mathbb{R}^2)$. Consequently, from (3.27), (3.37), (3.38), (3.40) and (3.12) we deduce that for $|x - y|$ small

$$\begin{aligned}D(x,y) = &-(2\pi)^{-1}\ln|x-y|(a'E_{\gamma\gamma} + \mu^{-1}E_{33}) \\ &+ 2a\mu(\lambda + \mu)(x_\alpha - y_\alpha)(x_\beta - y_\beta)|x-y|^{-2}E_{\alpha\beta} + \tilde{D}(x,y),\end{aligned} \quad (3.41)$$

$$\begin{aligned}P(x,y) = &-(2\pi)^{-1}\big[\mu'\varepsilon_{\alpha\beta}(\partial_{s(y)}\ln|x-y|)E_{\alpha\beta} + (\partial_{\nu(y)}\ln|x-y|)E_3 \\ &- (\lambda' + \mu')\varepsilon_{\alpha\gamma}\partial_{s(y)}\big((x_\alpha - y_\alpha)(x_\beta - y_\beta)|x-y|^{-2}\big)E_{\gamma\beta} \\ &+ \tfrac{1}{2}\varepsilon_{\alpha\beta}\partial_{s(y)}\big((x_\alpha - y_\alpha)\ln|x-y|\big)(\lambda'E_{3\beta} + h^{-2}E_{\beta 3}) \\ &- \tfrac{1}{2}\partial_{\nu(y)}\big((x_\alpha - y_\alpha)\ln|x-y|\big)E_{3\alpha}\big] + \tilde{P}(x,y),\end{aligned} \quad (3.42)$$

where

$$\begin{aligned}a' &= (\lambda + 3\mu)\big[2h^2\mu(\lambda + 2\mu)\big]^{-1}, \\ \lambda' &= \lambda(\lambda + 2\mu)^{-1}, \quad \mu' = \mu(\lambda + 2\mu)^{-1}\end{aligned} \quad (3.43)$$

and $\tilde{D}(x,y)$ and $\tilde{P}(x,y)$ satisfy the conditions in Theorem 1.3.

Before we establish the Somigliana formulae for plates, we need a more detailed analysis of the behaviour of $D(x,y)$ and $P(x,y)$ as one of the points x, y remains fixed while the other moves away from the origin. This analysis is based on a further refinement of (1.12), beyond the extra terms specified in (2.46). Thus, as $|x| \to \infty$,

$$\begin{aligned}|x-y|^{-2} = &\, |x|^{-2} + 2(x_1y_1 + x_2y_2)|x|^{-4} - |y|^2|x|^{-4} \\ &+ 4(x_1y_1 + x_2y_2)^2|x|^{-6} + O(|x|^{-5}), \\ \ln|x-y| = &\, \ln|x| - (x_1y_1 + x_2y_2)|x|^{-2} + \tfrac{1}{2}|y|^2|x|^{-2} \\ &- (x_1y_1 + x_2y_2)^2|x|^{-4} + (x_1y_1 + x_2y_2)|y|^2|x|^{-4} \\ &- \tfrac{4}{3}(x_1y_1 + x_2y_2)^3|x|^{-6} + O(|x|^{-4}).\end{aligned} \quad (3.44)$$

To this we adjoin the asymptotic formula [1]

$$K_0(h^{-1}|x-y|) = O(|x|^{-1/2}e^{-|x|}). \tag{3.45}$$

From (3.44), (3.45), (3.27), (3.37), (3.40), (3.12) and (2.26) we obtain the estimates

$$\begin{aligned}
D_{33} &= O(|x|^2 \ln|x|), \\
D_{\alpha 3}, D_{3\alpha} &= O(|x|\ln|x|), \\
D_{11}, D_{22} &= O(\ln|x|), \\
D_{12}, D_{21} &= O(1), \\
P_{3\alpha} &= O(\ln|x|), \\
P_{\alpha\beta}, P_{33} &= O(|x|^{-1}), \\
P_{\alpha 3} &= O(|x|^{-2}).
\end{aligned} \tag{3.46}$$

3.10. Theorem. (i) *If* $u \in C^2(S^+) \cap C^1(\bar{S}^+)$ *satisfies* $(Au)(x) = 0$, $x \in S^+$, *then*

$$\int_{\partial S} \left[D(x,y)(Tu)(y) - P(x,y)u(y)\right] ds(y) = \begin{cases} u(x) & x \in S^+, \\ \tfrac{1}{2}u(x) & x \in \partial S, \\ 0 & x \in S^-. \end{cases}$$

(ii) *If* $u \in C^2(S^-) \cap C^1(\bar{S}^-) \cap \mathcal{A}$ *satisfies* $(Au)(x) = 0$, $x \in S^-$, *then*

$$-\int_{\partial S} \left[D(x,y)(Tu)(y) - P(x,y)u(y)\right] ds(y) = \begin{cases} 0 & x \in S^+, \\ \tfrac{1}{2}u(x) & x \in \partial S, \\ u(x) & x \in S^-. \end{cases}$$

Proof. (i) This part is established in exactly the same way as Theorem 2.10(i), with the Greek indices replaced by Latin ones.

(ii) The equality

$$\left(-\int_{\partial S} + \int_{\partial K_R}\right)\left[D(x,y)(Tu)(y) - P(x,y)u(y)\right]ds(y) = u(x) \tag{3.47}$$

is also derived as in the proof of Theorem 2.10(ii), and we need to show that the integral over ∂K_R vanishes as $R \to \infty$. If we use (3.46) and the fact that, for $u \in \mathcal{A}$,

$$u_\alpha = O(R^{-1}), \quad u_3 = O(\ln R),$$

then we see that the integrand in (3.47) is $O(R^{-1}\ln R)$ and the required vanishing act does not seem to happen. Once again, we need to derive finer estimates by using the explicit expansions (3.19) and (3.20), and (3.23). Thus, after a simple but tedious computation, we arrive at the asymptotic equalities

$$D_{\alpha i}(x,y)(Tu)_i(y) - P_{\alpha i}(x,y)u_i(y) = O(R^{-2}\ln R),$$

$$D_{3\alpha}(x,y)(Tu)_\alpha(y) - P_{3i}(x,y)u_i(y)$$
$$= [4(\lambda+2\mu)]^{-1}(4\lambda\ln R + 3\lambda + 2\mu)[m_0\sin 2\theta + 2(m_2 - m_1)\cos 2\theta]R^{-1}$$
$$+ O(R^{-2}\ln R),$$

$$D_{33}(x,y)(Tu)_3(y) = [(m_7+m_9-2m_{11})\sin 2\theta + (m_8+m_{10}-2m_{12})\cos 2\theta]R^{-1}\ln R$$
$$+ O(R^{-2}\ln R).$$

This yields an expression of the form

$$D(x,y)(Tu)(y) - P(x,y)u(y) = \mathcal{G}(R,\theta) + O(R^{-2}\ln R),$$

where

$$\int_0^{2\pi} \mathcal{G}(R,\theta)\,d\theta = 0.$$

Consequently, the integrand in (3.47) is, in fact, $O(R^{-2}\ln R)$, which leads to the second Somigliana formula.

A similar analysis is made when $x \in \partial S$ and when $x \in S^+$.

3.4. Uniqueness theorems

These assertions, like many others in what follows, are handled in the same way as in Chapter 2. From now on we list all such statements without proof.

3.11. Theorem. (i) (D^+), (D^-), (N^-), (R^+) and (R^-) have at most one solution.
(ii) Any two solutions of (N^+) differ by a rigid displacement Fk.

3.12. Corollary. (i) If $u^\mathcal{A} + Fk$, where $k \in \mathcal{M}_{3\times 1}$ is constant, is a solution of the homogeneous problem (D^-) or (R^-), then

$$k = 0, \quad u^\mathcal{A} = 0.$$

(ii) If $u \in \mathcal{A}$ is a solution of (D^-) with $u|_{\partial S} = Fk$, where $k \in \mathcal{M}_{3\times 1}$ is constant, then

$$k = 0, \quad u = 0.$$

3.5. The plate potentials

Once again, these important tools in boundary integral equation methods are defined in terms of the matrices $D(x,y)$ and $P(x,y)$ constructed in §3.3.

3.13. Definition. The single layer and double layer plate potentials are defined by

$$(V\varphi)(x) = \int_{\partial S} D(x,y)\varphi(y)\,ds(y),$$

$$(W\psi)(x) = \int_{\partial S} P(x,y)\psi(y)\,ds(y),$$

where $\varphi, \psi \in \mathcal{M}_{3\times 1}$ are density functions required to satisfy certain smoothness properties.

3.14. Theorem. *If $\varphi, \psi \in C(\partial S)$, then*
(i) $V\varphi \in \mathcal{A}$ *if and only if* $p\varphi = 0$;
(ii) $W\psi \in \mathcal{A}$.

Proof. Substituting the expressions (3.44) in (3.27) and (3.40), we find that

$$\begin{aligned}(V\varphi)(r,\theta) &= M^\infty(r.\theta)(p\varphi) + (V\varphi)^{\mathcal{A}}(r,\theta),\\ (W\psi)(r,\theta) &= (W\psi)^{\mathcal{A}}(r,\theta),\end{aligned} \quad (3.48)$$

where $M^\infty(r,\theta)$ is the (3×3)-matrix with columns

$$M^{\infty(1)}(r,\theta) = -a\mu\big(\mu(2\ln r + 2 + \cos 2\theta), \mu \sin 2\theta,$$
$$- (\mu r(2\ln r + 1) - 4h^2(\lambda + 2\mu)r^{-1})\cos\theta\big)^{\mathrm{T}},$$
$$M^{\infty(2)}(r,\theta) = -a\mu\big(\sin 2\theta, \mu(2\ln r + 2 - \cos 2\theta),$$
$$- (\mu r(2\ln r + 1) - 4h^2(\lambda + 2\mu)r^{-1})\sin\theta\big)^{\mathrm{T}},$$
$$M^{\infty(3)}(r,\theta) = -a\mu\big(\mu r(2\ln r + 1)\cos\theta, \mu r(2\ln r + 1)\sin\theta,$$
$$- \mu r^2 \ln r + 4h^2(\lambda + 2\mu)\ln r + 4h^2(\lambda + 3\mu)\big)^{\mathrm{T}}.$$

The coefficients $m_0, \ldots, m_{12}$ with which $(V\varphi)^{\mathcal{A}}$ and $(W\psi)^{\mathcal{A}}$ fit the pattern (3.19) are too cumbersome and we omit their explicit expressions.

3.15. Lemma. $(AM^\infty)(x) = 0$, $x \in S^-$.

3.16. Theorem. (i) *If $\varphi, \psi \in C(\partial S)$, then $(V\varphi)(x)$ and $(W\psi)(x)$ are analytic at all $x \in S^+ \cup S^-$ and*

$$A(V\varphi)(x) = A(W\psi)(x) = 0, \quad x \in S^+ \cup S^-.$$

(ii) If $\varphi \in C^{0,\alpha}(\partial S)$, then the direct values $V_0\varphi$ and $W_0\psi$ of $V\varphi$ and $W\psi$ on ∂S exist (the latter in the sense of principal value). Also, the operators $\mathcal{V}^\pm$ defined by

$$\mathcal{V}^+\varphi = (V\varphi)|_{\bar{S}^+}, \quad \mathcal{V}^-\varphi = (V\varphi)|_{\bar{S}^-}$$

map $C^{0,\alpha}(\partial S)$ to $C^{1,\alpha}(\bar{S}^\pm)$, $\alpha \in (0,1)$, respectively, and

$$T(\mathcal{V}^+\varphi) = (W_0^* + \tfrac{1}{2}I)\varphi, \quad T(\mathcal{V}^-\varphi) = (W_0^* - \tfrac{1}{2}I)\varphi, \quad \varphi \in C^{0,\alpha}(\partial S),$$

where I is the identity operator and W_0^* is the adjoint of the direct value operator W_0, defined (in the sense of principal value) by

$$(W_0^*\varphi)(x) = \int_{\partial S} (T(\partial_x)D(x,y))\varphi(y)\,ds(y), \quad x \in \partial S.$$

(iii) The operators $\mathcal{W}^\pm$ defined by

$$\mathcal{W}^+\psi = \begin{cases} (W\psi)|_{S^+} & \text{in } S^+, \\ (W_0 - \tfrac{1}{2}I)\psi & \text{on } \partial S, \end{cases} \quad \mathcal{W}^-\psi = \begin{cases} (W\psi)|_{S^-} & \text{in } S^-, \\ (W_0 + \tfrac{1}{2}I)\psi & \text{on } \partial S \end{cases}$$

map $C^{0,\alpha}(\partial S)$ to $C^{0,\alpha}(\bar{S}^\pm)$ and $C^{1,\alpha}(\partial S)$ to $C^{1,\alpha}(\bar{S}^\pm)$, $\alpha \in (0,1)$, respectively, and

$$T(\mathcal{W}^+\psi) = T(\mathcal{W}^-\psi), \quad \psi \in C^{1,\alpha}(\partial S).$$

(iv) The operator W_0 maps $C^{0,\alpha}(\partial S)$ to $C^{1,\alpha}(\partial S)$, $\alpha \in (0,1)$.

The proof of this assertion follows from Theorems 1.3 and 1.4 since, by (3.41) and (3.42), we can write

$$V\varphi = v\big((a'E_{\gamma\gamma} + \mu^{-1}E_{33})\varphi\big) + 2a\mu(\lambda+\mu)v^b_{\alpha\beta}(E_{\alpha\beta}\varphi) + \tilde{V}\varphi,$$
$$W\psi = w\psi - (2\pi)^{-1}\big[\mu'v^f(\varepsilon_{\alpha\beta}E_{\alpha\beta}\psi) - (\lambda'+\mu')v^e_{\alpha\beta}(\varepsilon_{\alpha\gamma}E_{\gamma\beta}\psi)$$
$$+ \tfrac{1}{2}v^c_\alpha\big(\varepsilon_{\alpha\beta}(\lambda'E_{3\beta} + h^{-2}E_{\beta 3})\psi\big) - \tfrac{1}{2}v^d_\alpha(E_{3\alpha}\psi)\big] + \tilde{W}\psi,$$

where

$$(\tilde{V}\varphi)(x) = \int_{\partial S} \tilde{D}(x,y)\varphi(y)\,ds(y),$$
$$(\tilde{W}\psi)(x) = \int_{\partial S} \tilde{P}(x,y)\psi(y)\,ds(y).$$

Full details can be found in [3].

3.17. Remarks. (i) The derivatives on ∂S of the functions defined in $\bar{S}^+$ or $\bar{S}^-$ in Theorem 3.16 are one-sided.

(ii) $A(\mathcal{V}^+\varphi)(x) = 0$, $x \in S^+$, and $A(\mathcal{V}^-\varphi)(x) = 0$, $x \in S^-$.

(iii) If $\mathcal{V}^+\varphi = \mathcal{V}^-\varphi = 0$, then, by Theorem 3.16(ii), $\varphi = 0$.

(iv) With the notation in Theorem 3.16, we can rewrite the Somigliana relations in Theorem 3.10 as

$$\begin{aligned}\mathcal{V}^+(Tu) - \mathcal{W}^+(u|_{\partial S}) &= u, \\ \mathcal{V}^-(Tu) - \mathcal{W}^-(u|_{\partial S}) &= 0\end{aligned} \quad (3.49)$$

and

$$\begin{aligned}-\mathcal{V}^-(Tu) + \mathcal{W}^-(u|_{\partial S}) &= u, \\ -\mathcal{V}^+(Tu) + \mathcal{W}^+(u|_{\partial S}) &= 0.\end{aligned} \quad (3.50)$$

(v) From (3.49) with $u = F^{(i)}$ we derive the equalities

$$\mathcal{W}^+ F = -F,$$
$$W_0 F = -\tfrac{1}{2}F,$$
$$\mathcal{W}^- F = 0.$$

They do not contradict (3.50) since the latter are valid for $u \in \mathcal{A}$, and $F \notin \mathcal{A}$.

(vi) The last formula in Theorem 3.16(iii) enables us to define a boundary operator

$$N_0 : C^{1,\alpha}(\partial S) \to C^{0,\alpha}(\partial S)$$

by writing

$$N_0\psi = T(\mathcal{W}^+\psi) = T(\mathcal{W}^-\psi), \quad \psi \in C^{1,\alpha}(\partial S).$$

In view of Theorem 3.16, we may use the notation

$$(\mathcal{V}^\pm\varphi)|_{\partial S} = \mathcal{V}_0^\pm\varphi = V_0\varphi,$$
$$(\mathcal{W}^\pm\psi)|_{\partial S} = \mathcal{W}_0^\pm\psi = (W_0 \mp \tfrac{1}{2}I)\psi.$$

From now on in this chapter we assume that the boundary integral operators V_0, W_0 and W_0^* and the combinations $W_0 \pm \tfrac{1}{2}I$ and $W_0^* \pm \tfrac{1}{2}I$ are defined on $C^{0,\alpha}(\partial S)$, while N_0 is defined on $C^{1,\alpha}(\partial S)$, $\alpha \in (0,1)$.

3.6. Properties of the boundary operators

The algebra of these operators works just as in Chapter 2.

3.18. Theorem. V_0, W_0, W_0^* and N_0 satisfy the composition formulae

$$W_0 V_0 = V_0 W_0^*, \quad N_0 V_0 = W_0^{*2} - \tfrac{1}{4}I \quad \text{on } C^{0,\alpha}(\partial S),$$
$$N_0 W_0 = W_0^* N_0, \quad V_0 N_0 = W_0^2 - \tfrac{1}{4}I \quad \text{on } C^{1,\alpha}(\partial S).$$

3.19. Theorem. $W_0 \pm \tfrac{1}{2}I$ and $W_0^* \pm \tfrac{1}{2}I$ are operators of index zero.

Proof. Proceeding as in the case of Theorem 2.19 and using (3.41) and (3.42), we can write

$$(W_0 - \tfrac{1}{2}I)\varphi = K^s \varphi + K^w \varphi - \tfrac{1}{2}\varphi,$$

where

$$(K^s \varphi)(z) = -(2\pi)^{-1} \mu' \varepsilon_{\alpha\beta} E_{\alpha\beta} \int_{\partial S} (\zeta - z)^{-1} \varphi(\zeta)\, d\zeta, \quad z \in \partial S,$$

and K^w is a weakly singular operator. We see that Definition 1.16 holds with

$$\hat{k}^s(z,z) = -(2\pi)^{-1} \mu' \varepsilon_{\alpha\beta} E_{\alpha\beta}, \quad z \in \partial S,$$

while, as before, $\hat{k}^w(z,z) = 0$. Hence, $W_0 - \tfrac{1}{2}I$ is α-regular singular, $\alpha \in (0,1)$, and

$$\hat{k}(z,z) = \hat{k}^s(z,z) + \hat{k}^w(z,z) = -(2\pi)^{-1} \mu' \varepsilon_{\alpha\beta} E_{\alpha\beta}, \quad z \in \partial S;$$

therefore, by (3.43),

$$\det\left[-\tfrac{1}{2}E_3 \pm \pi i \hat{k}(z,z)\right] = -\tfrac{1}{8}(1 - \mu'^2)$$
$$= -\tfrac{1}{8}(\lambda + \mu)(\lambda + 3\mu)(\lambda + 2\mu)^{-2} \neq 0, \quad z \in \partial S,$$

which means that the index of $W_0 - \tfrac{1}{2}I$ (see (1.4)) is zero.

We show in a similar manner that the index of $W_0 + \tfrac{1}{2}I$ is zero. Finally, by Remark 1.18(ii), $W_0^* \pm \tfrac{1}{2}I$ are also operators of index zero.

3.20. Corollary. *The Fredholm Alternative holds for the pairs of equations*

$$(W_0 - \tfrac{1}{2}I)\varphi = f, \quad (W_0^* - \tfrac{1}{2}I)\psi = g, \quad f, g \in C^{0,\alpha}(\partial S),$$
$$(W_0 + \tfrac{1}{2}I)\varphi = f, \quad (W_0^* + \tfrac{1}{2}I)\psi = g, \quad f, g \in C^{0,\alpha}(\partial S),$$

in the dual system $\bigl(C^{0,\alpha}(\partial S), C^{0,\alpha}(\partial S)\bigr)$, $\alpha \in (0,1)$, with the bilinear form

$$(\varphi, \psi) = \int_{\partial S} \varphi^{\mathrm{T}} \psi \, ds.$$

3.21. Theorem. *If there is $\varphi \in C^{0,\alpha}(\partial S)$, $\varphi \neq 0$, such that*

$$(W_0^* + \tfrac{1}{2}I)\varphi = 0,$$

then

$$\mathcal{V}^+ \varphi = Fk$$

for some constant $k \in \mathcal{M}_{3\times 1}$, and

$$p\varphi \neq 0.$$

3.22. Theorem. (i) *The null spaces of $W_0 - \tfrac{1}{2}I$ and $W_0^* - \tfrac{1}{2}I$ consist of the zero vector alone.*

(ii) *The null spaces of $W_0 + \tfrac{1}{2}I$ and $W_0^* + \tfrac{1}{2}I$ are three-dimensional and are spanned, respectively, by $\{F^{(i)}\}$ and the (linearly independent) columns $\{\Phi^{(i)}\}$ of a (3×3)-matrix $\Phi \in C^{0,\alpha}(\partial S)$.*

Proof. This is conducted just as in the case of Theorem 2.22 except that here we replace (2.65) by

$$c = -(\mathcal{W}^+ f^{(0)})(0,0).$$

With this choice, $k = 0$ and then $f^{(0)} = Fc$, as required.

3.23. Remark. Φ is not unique.

3.24. Theorem. (i) *The null spaces of $W_0^2 - \tfrac{1}{4}I$ and $W_0^{*2} - \tfrac{1}{4}I$ coincide with those of $W_0 + \tfrac{1}{2}I$ and $W_0^* + \tfrac{1}{2}I$ (that is, they are spanned by $\{F^{(i)}\}$ and $\{\Phi^{(i)}\}$, respectively).*

(ii) *$N_0 \psi = 0$ if and only if $\psi = Fk$, where $k \in \mathcal{M}_{3\times 1}$ is constant.*

3.25. Theorem. *For every simple closed C^2-curve ∂S and any $\alpha \in (0,1)$, there are a unique (3×3)-matrix $\Phi \in C^{0,\alpha}(\partial S)$ and a unique constant $\mathcal{C} \in \mathcal{M}_{3\times 3}$ such that the columns $\Phi^{(i)}$ of Φ are linearly independent and*

$$V_0 \Phi = F\mathcal{C}, \quad p\Phi = E_3.$$

3.26. Remarks. (i) The equality $p\Phi = E_3$ is equivalent to the biorthogonality of the sets $\{F^{(i)}\}$ and $\{\Phi^{(j)}\}$; that is,

$$(F^{(i)}, \Phi^{(j)}) = \int_{\partial S} (F^{(i)})^{\mathrm{T}} \Phi^{(j)}\, ds = \delta_{ij}, \quad i,j = 1,2,3.$$

(ii) Since (D^+) has at most one solution, it follows that $\mathcal{V}^+\Phi = F\mathcal{C}$.

(iii) No obvious connection can be found between the Robin constant (or the logarithmic capacity of ∂S) and the matrix $\mathcal{C}$.

3.27. Remark. Since the $D_{\alpha\beta}$ contain K_0 explicitly, integrals involving their expressions are difficult to compute, even when ∂S is a circle, so we do not have a readily available example that produces the matrices Φ and $\mathcal{C}$ for some specific boundary curve. This will be partially remedied later (see Example 3.31).

3.28. Corollary. *If $V_0\varphi = Fc$ with a constant $c \in \mathcal{M}_{3\times 1}$, then*

$$\varphi = \Phi k, \quad c = \mathcal{C}k$$

for some constant $k \in \mathcal{M}_{3\times 1}$.

3.29. Theorem. *The equation $V_0\varphi = 0$ has non-zero solutions if and only if ∂S is such that $\det \mathcal{C} = 0$.*

3.30. Corollary. *The null space of V_0 is the subspace of the null space of $W_0^* + \tfrac{1}{2}I$ consisting of all functions of the form $\varphi = \Phi k$, where k is any constant (3×1)-vector such that $\mathcal{C}k = 0$. More precisely, if $\operatorname{rank} \mathcal{C} = i$, $i = 0,1,2,3$, then the dimension of the null space of V_0 is $3-i$. In particular, if $\det \mathcal{C} \neq 0$, this space consists of the zero vector alone; if $\mathcal{C} = 0$, it coincides with the null space of $W_0^* + \tfrac{1}{2}I$.*

3.31. Example. It is clear that the matrix $D(x,y)$ of fundamental solutions is not unique: we can add to it any (3×3)-matrix $\hat{D}(x,y)$ such that

$$A(\partial_x)\hat{D}(x,y) = 0, \quad x,y \in \mathbb{R}^2,$$

provided that $(D + \hat{D})(x,y)$ satisfies (3.39); in other words,

$$(D + \hat{D})(x,y) = (D + \hat{D})^{\mathrm{T}}(y,x).$$

We choose $\hat{D}$ to be the constant matrix whose entries are all equal to zero except $\hat{D}_{33}$, which is equal to $a\mu c$, $c = \text{const}$. For simplicity, we continue to use the old symbol D instead of $D + \hat{D}$.

Let ∂S be the circle ∂K_R with the centre at the origin and radius R. Direct computation shows that for all $x \in \partial K_R$

$$\int_{\partial K_R} \ln|x-y|\, ds(y) = 2\pi R \ln R,$$

$$\int_{\partial K_R} |x-y|^2 \ln|x-y|\, ds(y) = 2\pi R^3(2\ln R + 1),$$

$$\int_{\partial K_R} y_\alpha \ln|x-y|\, ds(y) = -\pi R x_\alpha;$$

therefore, by (3.27), (3.38)$_3$, (3.38)$_2$ (with an additional term $a\mu c$) and (3.37),

$$\int_{\partial K_R} D_{\alpha 3}(x,y)\, ds(y) = \int_{\partial K_R} (\operatorname{adj} A)_{\alpha 3}(\partial_x) t(x,y)\, ds(y) = -4\pi a\mu^2 R(\ln R + 1) x_\alpha,$$

$$\int_{\partial K_R} D_{33}(x,y)\, ds(y) = \int_{\partial K_R} (\operatorname{adj} A)_{33}(\partial_x) t(x,y)\, ds(y) \qquad (3.51)$$

$$= 2\pi a\mu R\{[\mu R^2 - 4h^2(\lambda + 2\mu)](\ln R + 1) + \mu R^2 \ln R - 4h^2\mu + c\}.$$

If we take

$$R = e^{-1}, \quad c = \mu(e^{-2} + 4h^2), \qquad (3.52)$$

then equalities (3.51) become

$$\int_{\partial K_R} D_{i3}(x,y)\, ds(y) = 0,$$

and we can write

$$V_0 F^{(3)} = 0.$$

This means that for the choice (3.52) the null space of V_0 contains non-zero vectors. By Corollary 3.28, there is a constant $k \in \mathcal{M}_{3\times 1}$ such that

$$\Phi k = F^{(3)}, \quad \mathcal{C}k = 0.$$

Since $F^{(3)} \neq 0$, it follows that $k \neq 0$, so $\mathcal{C}$ is singular, as predicted by Theorem 3.29.

3.32. Lemma. (i) *The characteristic matrix $\mathcal{C}$ is symmetric; that is,*

$$\mathcal{C}^{\mathrm{T}} = \mathcal{C},$$

and if q is the operator defined on $C(\partial S)$ by

$$q\varphi = \int_{\partial S} \Phi^{\mathrm{T}}\varphi\, ds,$$

then for any $\varphi \in C(\partial S)$

$$q(V_0\varphi) = \mathcal{C}(p\varphi).$$

(ii) *For any* $\psi \in C^{0,\alpha}(\partial S)$

$$p(W_0^*\psi) = -\tfrac{1}{2}p\psi.$$

(iii) *For any* $\psi \in C^{0,\alpha}(\partial S)$

$$q(W_0\psi) = -\tfrac{1}{2}q\psi.$$

(iv) *For any* $\psi \in C^{1,\alpha}(\partial S)$

$$p(N_0\psi) = 0.$$

3.7. Boundary integral equation methods

In §§3.1–3.6 we have developed all the background details required by the construction of boundary integral equation methods in the case of equilibrium bending of plates with transverse shear deformation. It turns out that, once this information is available, the development of the techniques discussed in §§2.7–2.12, both direct and indirect, follows exactly the same pattern here as for plane strain. The various symbols now stand for different quantities, but the mathematical procedures, relationships and assertions remain the same and could be reproduced verbatim. For this reason we omit them all. However, to help the reader follow these methods with direct reference to plates, we have arranged that, with one (inconsequential) exception, every single subheading in the first 6 sections of this chapter mirrors its counterpart in Chapter 2.

Before we conclude Chapter 3, we clarify the statement in Remark 3.5 by indicating exactly how the boundary value problems (D$^\pm$), (N$^\pm$) and (R$^\pm$) for the non-homogeneous system (3.9); that is,

$$(Au)(x) = f(x), \quad x \in S^\pm, \tag{3.53}$$

where f is a prescribed function, can be reduced to the corresponding problems for (3.9), stated in Definition 3.4. We consider (D$^+$) as an example.

In [23] it is shown that if $f \in C^{1,\alpha}(S^+)$, $\alpha \in (0,1)$, then the function

$$U(x) = \int_{S^+} D(x,y) f(y) \, da(y), \quad x \in S^+, \tag{3.54}$$

is a (regular) solution of (3.53) in S^+. Furthermore, in [24] it is proved that if $f \in L^\infty(S^+)$, then $U(x)$ defined by (3.54) with $x \in \partial S$ belongs to $C^{1,\alpha}(\partial S)$, $\alpha \in (0,1)$. This means that for $f \in C^{1,\alpha}(S^+) \cap L^\infty(S^+)$ the function

$$\bar{u}(x) = u(x) - U(x), \quad x \in \bar{S}^+,$$

where u is the solution of (3.53) in S^+ such that $u|_{\partial S} = \mathcal{P}$, satisfies

$$\begin{aligned}(A\bar{u})(x) &= 0, \quad x \in S^+, \\ \bar{u}(x) &= \mathcal{P}(x) - U(x), \quad x \in \partial S,\end{aligned} \tag{3.55}$$

which is (D$^+$) for (3.9). Since in the boundary integral equations for (D$^+$) we require $\mathcal{P} \in C^{1,\alpha}(\partial S)$, we see that this condition is also satisfied by the new boundary data in (3.55).

The other boundary value problems are treated similarly, with some additional restrictions on the behaviour of f in the far field for the exterior ones.

Chapter 4
Which Method?

4.1. Notation and prerequisites

As we have seen, there are essentially two distinct categories of boundary integral equation methods: direct and indirect ones. The construction of these two types is based on different approaches. The former start from the representation formulae, where the solution at any point in the domain is expressed in terms of its boundary value and its image under the associated boundary operator. If the problem models a physical situation, then the unknown functions have a direct physical significance. By contrast, the indirect methods use "designer" solutions, postulated in a form that makes use of an abstract function chosen exclusively on grounds of mathematical convenience.

Since the original problem is the same in both cases, it is obvious that, whichever representation we choose, it must ultimately produce the same solution. It is, therefore, interesting and useful to compare these representations, find the mathematical connections between the various unknown functions and decide which of the techniques seems to be best suited for computational work.

In what follows we quote the necessary results from Chapter 2. However, the discussion is general and its conclusions for the Laplace equation and bending of elastic plates are exactly the same, as is the notation.

Before we begin, we compile a list of the boundary integral equations established and solved in each method, using different symbols for the different density functions involved. For simplicity, we confine ourselves to the interior Dirichlet problem (D^+).

Thus, the integral equations in §§2.7–2.12 are now written in the form

$$(W_0 - \tfrac{1}{2}I)\varphi = \mathcal{P}, \qquad (\mathcal{D}_C^+)$$

$$V_0 \theta = \mathcal{P}, \qquad (\mathcal{D}_A^+)$$

$$V_0 \chi + (FH)(p\chi) = \mathcal{P}, \qquad (\mathcal{D}_M^+)$$

$$V_0 \sigma - Fc = \mathcal{P}, \quad p\sigma = s, \qquad (\mathcal{D}_R^+)$$

$$V_0 \psi = (W_0 + \tfrac{1}{2}I)\mathcal{P}, \quad p\psi = 0, \qquad (\mathcal{D}_D^+)$$

$$(W_0^* - \tfrac{1}{2}I)\tau = N_0 \mathcal{P}. \qquad (\mathcal{D}_S^+)$$

Throughout this chapter we assume that the Dirichlet boundary data $\mathcal{P}$ is a $C^{1,\alpha}$-function.

4.2. Connections between the indirect methods

We start by looking at connections between the solutions of the integral equations arising from the classical, alternative, modified and refined indirect methods.

(i) $(\mathcal{D}_\mathrm{C}^+)$ and $(\mathcal{D}_\mathrm{A}^+)$. Since, by Theorem 2.33(i), $(\mathcal{D}_\mathrm{C}^+)$ has a unique solution, the operator $W_0 - \tfrac{1}{2}I : C^{1,\alpha}(\partial S) \to C^{1,\alpha}(\partial S)$ is invertible and we can write

$$\varphi = (W_0 - \tfrac{1}{2}I)^{-1}\mathcal{P} = \big((W_0 - \tfrac{1}{2}I)^{-1}V_0\big)\theta.$$

Conversely, if $\det \mathcal{C} \neq 0$, then, by Corollary 2.30, $V_0 : C^{0,\alpha}(\partial S) \to C^{1,\alpha}(\partial S)$ is invertible, so

$$\theta = V_0^{-1}\mathcal{P} = \big(V_0^{-1}(W_0 - \tfrac{1}{2}I)\big)\varphi.$$

(ii) $(\mathcal{D}_\mathrm{C}^+)$ and $(\mathcal{D}_\mathrm{M}^+)$. As in (i),

$$\varphi = (W_0 - \tfrac{1}{2}I)^{-1}\mathcal{P} = (W_0 - \tfrac{1}{2}I)^{-1}\big(V_0\chi + (FH)(p\chi)\big).$$

On the other hand, $(\mathcal{D}_\mathrm{M}^+)$ is, in fact, the equation (see §2.9)

$$V_0^H \chi = \mathcal{P},$$

which, by Theorem 2.42, is uniquely solvable; therefore, $V_0^H : C^{0,\alpha}(\partial S) \to C^{1,\alpha}(\partial S)$ is invertible and we find that

$$\chi = (V_0^H)^{-1}\mathcal{P} = (V_0^H)^{-1}(W_0 - \tfrac{1}{2}I)\varphi.$$

(iii) $(\mathcal{D}_\mathrm{C}^+)$ and $(\mathcal{D}_\mathrm{R}^+)$. Again as in (i),

$$\varphi = (W_0 - \tfrac{1}{2}I)^{-1}\mathcal{P} = (W_0 - \tfrac{1}{2}I)^{-1}(V_0\sigma - Fc).$$

The other way round, by (2.108) and Lemma 2.32(iii),

$$\begin{aligned}V_0\sigma &= \mathcal{P} + Fc = \mathcal{P} + F(\mathcal{C}s - q\mathcal{P}) \\ &= (W_0 - \tfrac{1}{2}I)\varphi + F\big(\mathcal{C}s - q(W_0 - \tfrac{1}{2}I)\varphi\big) \\ &= (W_0 - \tfrac{1}{2}I)\varphi + F(\mathcal{C}s + q\varphi).\end{aligned}$$

If $\det \mathcal{C} \neq 0$, then, by Corollary 2.30,

$$\sigma = V_0^{-1}\big((W_0 - \tfrac{1}{2}I)\varphi + F(\mathcal{C}s + q\varphi)\big).$$

(iv) $(\mathcal{D}_A^+)$ *and* $(\mathcal{D}_M^+)$. Subtracting the two equations term by term, we find that

$$V_0(\theta - \chi) = (FH)(p\chi) = F\big(H(p\chi)\big),$$

so, by Corollary 2.28,

$$\theta - \chi = \Phi k, \quad H(p\chi) = \mathcal{C}k \tag{4.1}$$

for some constant $k \in \mathcal{M}_{3\times 1}$. Hence, since $p\Phi = E_3$,

$$p\theta - p\chi = (p\Phi)k = k, \tag{4.2}$$

and from $(4.1)_2$ it follows that

$$H(p\chi) = \mathcal{C}k = \mathcal{C}(p\theta - p\chi).$$

Recalling that H is chosen so that $H + \mathcal{C}$ is invertible, this yields

$$p\chi = (H + \mathcal{C})^{-1}\mathcal{C}(p\theta), \tag{4.3}$$

and $(4.1)_1$, (4.2) and (4.3) imply that

$$\chi = \theta - \Phi k = \theta - \Phi(p\theta - p\chi) = \theta - \Phi\big(p\theta - (H + \mathcal{C})^{-1}\mathcal{C}(p\theta)\big). \tag{4.4}$$

If $\det \theta = 0$, then θ is not uniquely determined; according to (4.4), it would seem that neither is χ, which would contradict Theorem 2.42. To resolve this apparent paradox, let $\bar\chi$ be constructed by means of (4.4) from another solution $\bar\theta \neq \theta$ of $(\mathcal{D}_A^+)$. Then

$$V_0(\theta - \bar\theta) = 0, \tag{4.5}$$

which, by Corollary 2.28, means that

$$\theta - \bar\theta = \Phi a, \quad \mathcal{C}a = 0 \tag{4.6}$$

for some constant $a \in \mathcal{M}_{3\times 1}$. From $(4.6)_1$ we also find that

$$p(\theta - \bar\theta) = (p\Phi)a = a; \tag{4.7}$$

therefore, by (4.4) and (4.6),

$$\chi - \bar{\chi} = \theta - \bar{\theta} - \Phi(p(\theta - \bar{\theta}) - (H+C)^{-1}Cp(\theta - \bar{\theta}))$$
$$= \Phi(a - a + (H+C)^{-1}(Ca)) = 0;$$

in other words, χ is indeed unique.

If $\det C \neq 0$, then θ is unique and from $(4.1)_2$ it follows that

$$k = (C^{-1}H)(p\chi),$$

so, by $(4.1)_1$,

$$\theta = \chi + (\Phi C^{-1}H)(p\chi).$$

If $\det C = 0$, then

$$\theta = \chi + \Phi k,$$

where k is any solution of $(4.1)_2$.

(v) $(\mathcal{D}_A^+)$ and $(\mathcal{D}_R^+)$. Subtracting the former from $(\mathcal{D}_R^+)_1$, we obtain

$$V_0(\sigma - \theta) = Fc,$$

so, by Corollary 2.28, there is some constant $k \in \mathcal{M}_{3\times 1}$ such that

$$\sigma - \theta = \Phi k, \quad c = Ck. \tag{4.8}$$

Consequently,

$$p\sigma - p\theta = (p\Phi)k = k,$$

which, since $p\sigma = s$, can be written as

$$k = s - p\theta.$$

By (4.8), this leads to

$$\sigma = \theta + \Phi(s - p\theta), \quad c = C(s - p\theta). \tag{4.9}$$

As remarked in (iv), θ may not be unique. Let $\bar{\sigma}, \bar{c}$ be another pair generated by (4.9) from a different solution $\bar{\theta} \neq \theta$ of $(\mathcal{D}_A^+)$. Then (4.5), (4.6) and (4.7) again hold, and (4.9) yields

$$\sigma - \bar{\sigma} = \theta - \bar{\theta} - \Phi p(\theta - \bar{\theta}) = \Phi a - \Phi a = 0,$$
$$c - \bar{c} = -Cp(\theta - \bar{\theta}) = -Ca = 0;$$

therefore, the pair σ, c is unique, as stipulated by Theorem 2.45.

Conversely, if $\det \mathcal{C} \neq 0$, then θ is unique and, by $(4.8)_2$,

$$k = \mathcal{C}^{-1}c,$$

so, according to $(4.8)_1$,

$$\theta = \sigma - \Phi k = \sigma - (\Phi \mathcal{C}^{-1})c.$$

If $\det \mathcal{C} = 0$, then

$$\theta = \sigma - \Phi k,$$

where k is any solution of $(4.8)_2$.

(vi) $(\mathcal{D}_M^+)$ and $(\mathcal{D}_R^+)$. We subtract $(\mathcal{D}_M^+)$ from $(\mathcal{D}_R^+)_1$ and arrive at

$$V_0(\sigma - \chi) = F(c + H(p\chi)).$$

By Corollary 2.28, there is a constant $k \in \mathcal{M}_{3\times 1}$ such that

$$\sigma - \chi = \Phi k, \quad c + H(p\chi) = \mathcal{C}k.$$

Then

$$p\sigma - p\chi = (p\Phi)k = k$$

and, by $(\mathcal{D}_R^+)_2$,

$$c = -H(p\chi) + \mathcal{C}k = -H(p\chi) + \mathcal{C}(p\sigma - p\chi) = \mathcal{C}(p\sigma) - (H + \mathcal{C})(p\chi),$$
$$\sigma = \chi + \Phi k = \chi + \Phi(p\sigma - p\chi),$$

which, by $(\mathcal{D}_R^+)_2$, become

$$c = \mathcal{C}s - (H + \mathcal{C})(p\chi), \tag{4.10}$$
$$\sigma = \chi + \Phi(s - p\chi). \tag{4.11}$$

The other way round, from (4.11) it follows that

$$\chi = \sigma - \Phi k = \sigma - \Phi(s - p\chi),$$

while (4.10) yields

$$p\chi = (H + \mathcal{C})^{-1}(\mathcal{C}s - c);$$

hence,

$$\chi = \sigma - \Phi\big(s - (H + \mathcal{C})^{-1}(\mathcal{C}s - c)\big).$$

4.3. Connections between the direct and indirect methods

Owing to the boundary operators involved, in this case the connections between the solutions of the direct and indirect methods are one-sided.

(i) $(\mathcal{D}_D^+)$ and $(\mathcal{D}_C^+)$. Applying $W_0 + \frac{1}{2}I$ to $(\mathcal{D}_C^+)$, we find that

$$f(W_0^2 - \tfrac{1}{4}I)\varphi = (W_0 + \tfrac{1}{2}I)\mathcal{P} = V_0\psi,$$

or, by $(4.1.2)_2$,

$$(V_0 N_0)\varphi = V_0\psi,$$

which we rewrite as

$$V_0(\psi - N_0\varphi) = 0.$$

If $\det \mathcal{C} \neq 0$, then, by Corollary 2.30,

$$\psi = N_0\varphi. \tag{4.12}$$

In view of Lemma 2.32(iv), ψ satisfies $(\mathcal{D}_D^+)_2$ automatically.

If $\det \mathcal{C} = 0$, then, again by Corollary 2.30,

$$\psi = N_0\varphi + \Phi a,$$

where $a \in \mathcal{M}_{3 \times 1}$ is any constant vector satisfying $\mathcal{C}a = 0$. In this case, $(\mathcal{D}_D^+)_2$ implies that we must have

$$(pN_0)\varphi + (p\Phi)a = 0,$$

which yields $a = 0$. Consequently, (4.12) holds whether $\mathcal{C}$ is singular or not.

(ii) $(\mathcal{D}_D^+)$ and $(\mathcal{D}_A^+)$. Applying $W_0 + \frac{1}{2}I$ to $(\mathcal{D}_A^+)$, we find that

$$\left((W_0 + \tfrac{1}{2}I)V_0\right)\theta = (W_0 + \tfrac{1}{2}I)\mathcal{P} = V_0\psi,$$

which, in view of $(4.1.1)_1$, is equivalent to

$$V_0\left(\psi - (W_0^* + \tfrac{1}{2}I)\theta\right) = 0.$$

If $\det \mathcal{C} \neq 0$, then, by Corollary 2.30,

$$\psi = (W_0^* + \tfrac{1}{2}I)\theta. \tag{4.13}$$

By Lemma 2.32(ii),
$$p\psi = p(W_0^*\theta) + \tfrac{1}{2}(p\theta) = 0,$$
so $(\mathcal{D}_D^+)_2$ is satisfied.

If $\det \mathcal{C} = 0$, then
$$\psi = (W_0^* + \tfrac{1}{2}I)\theta + \Phi a,$$
where a is any constant (3×1)-vector such that $\mathcal{C}a = 0$. For ψ to satisfy $(\mathcal{D}_D^+)_2$, we now must have
$$p\psi = p(W_0^*\theta) + \tfrac{1}{2}(p\theta) + (p\Phi)a = a = 0,$$
so (4.13) holds in either case.

(iii) $(\mathcal{D}_D^+)$ *and* $(\mathcal{D}_M^+)$. We operate with $W_0 + \tfrac{1}{2}I$ in $(\mathcal{D}_M^+)$ and see that
$$\left((W_0 + \tfrac{1}{2}I)V_0\right)\chi + (W_0 + \tfrac{1}{2}I)(FH)(p\varphi) = (W_0 + \tfrac{1}{2}I)\mathcal{P} = V_0\psi.$$

Since, by Remark 2.17(v), we have $W_0 F = -\tfrac{1}{2}F$, we use $(4.1.1)_1$ to rewrite this equation as
$$V_0\left(\psi - (W_0^* + \tfrac{1}{2}I)\chi\right) = 0.$$

The discussion now proceeds as in (ii) and leads to the equality
$$\psi = (W_0^* + \tfrac{1}{2}I)\chi.$$

(iv) $(\mathcal{D}_D^+)$ *and* $(\mathcal{D}_R^+)$. We use the same procedure as in (ii) and (iii) to obtain
$$\left((W_0 + \tfrac{1}{2}I)V_0\right)\sigma - (W_0 + \tfrac{1}{2}I)(Fc) = (W_0 + \tfrac{1}{2}I)\mathcal{P} = V_0\psi,$$
and then
$$V_0\left(\psi - (W_0^* + \tfrac{1}{2}I)\sigma\right) = 0,$$
from which it follows that
$$\psi = (W_0^* + \tfrac{1}{2}I)\sigma.$$

(v) $(\mathcal{D}_S^+)$ *and* $(\mathcal{D}_C^+)$, $(\mathcal{D}_A^+)$, $(\mathcal{D}_M^+)$, $(\mathcal{D}_R^+)$. The relationships are obviously the same as in (i)–(iv) since
$$\tau = Tu = \psi.$$

4.4. Overall view and conclusions

From the formulae established in §§4.2 and 4.3 it is clear that the solutions of the boundary integral equations in any of the indirect methods can parametrize those in all the other methods. It is also clear that none of the solutions of $(\mathcal{D}_C^+)$, $(\mathcal{D}_A^+)$, $(\mathcal{D}_M^+)$ and $(\mathcal{D}_R^+)$ can be expressed in terms of the "physical" solutions of $(\mathcal{D}_D^+)$. For convenience, we summarize all the connection formulae in one table.

$(\mathcal{D}_C^+)$ and $(\mathcal{D}_A^+)$: $\quad \varphi = ((W_0 - \tfrac{1}{2}I)^{-1} V_0) \theta$

$\qquad\qquad\qquad\qquad \theta = (V_0^{-1}(W_0 - \tfrac{1}{2}I))\varphi \qquad$ if $\det \mathcal{C} \neq 0$

$(\mathcal{D}_C^+)$ and $(\mathcal{D}_M^+)$: $\quad \varphi = (W_0 - \tfrac{1}{2}I)^{-1}(V_0 \chi + (FH)(p\chi))$

$\qquad\qquad\qquad\qquad \chi = (V_0^H)^{-1}(W_0 - \tfrac{1}{2}I)\varphi$

$(\mathcal{D}_C^+)$ and $(\mathcal{D}_R^+)$: $\quad \varphi = (W_0 - \tfrac{1}{2}I)^{-1}(V_0 \sigma - Fc)$

$\qquad\qquad\qquad\qquad \sigma = V_0^{-1}((W_0 - \tfrac{1}{2}I)\varphi + F(\mathcal{C}s + q\varphi)) \qquad$ if $\det \mathcal{C} \neq 0$

$(\mathcal{D}_A^+)$ and $(\mathcal{D}_M^+)$: $\quad \theta = \chi + (\Phi \mathcal{C}^{-1} H)(p\chi) \qquad$ if $\det \mathcal{C} \neq 0$

$\qquad\qquad\qquad\qquad \chi = \theta - \Phi(p\theta - (H + \mathcal{C})^{-1}\mathcal{C}(p\theta))$

$(\mathcal{D}_A^+)$ and $(\mathcal{D}_R^+)$: $\quad \theta = \sigma - (\Phi \mathcal{C}^{-1})c \qquad$ if $\det \mathcal{C} \neq 0$

$\qquad\qquad\qquad\qquad \sigma = \theta + \Phi(s - p\theta), \; c = \mathcal{C}(s - p\theta)$

$(\mathcal{D}_M^+)$ and $(\mathcal{D}_R^+)$: $\quad \chi = \sigma - \Phi(s - (H + \mathcal{C})^{-1}(\mathcal{C}s - c))$

$\qquad\qquad\qquad\qquad \sigma = \chi + \Phi(s - p\chi), \; c = \mathcal{C}s - (H + \mathcal{C})(p\chi)$

$(\mathcal{D}_D^+)$ and $(\mathcal{D}_C^+)$: $\quad \psi = N_0 \varphi$

$(\mathcal{D}_D^+)$ and $(\mathcal{D}_A^+)$: $\quad \psi = (W_0^* + \tfrac{1}{2}I)\theta$

$(\mathcal{D}_D^+)$ and $(\mathcal{D}_M^+)$: $\quad \psi = (W_0^* + \tfrac{1}{2}I)\chi$

$(\mathcal{D}_D^+)$ and $(\mathcal{D}_R^+)$: $\quad \psi = (W_0^* + \tfrac{1}{2}I)\sigma$

$(\mathcal{D}_D^+)$ and $(\mathcal{D}_S^+)$: $\quad \psi = \tau$

A very important point about all these methods is to identify which one is the most useful for numerical computations. Two features are essential in making such a decision: the corresponding equation should preferably be weakly singular (as opposed to strongly singular), it should be uniquely solvable, and its solution should not require foreknowledge of anything other than the prescribed data. The following

OVERALL VIEW AND CONCLUSIONS

summary encapsulates these features for each of the techniques included in the above table.

$(\mathcal{D}_C^+)$ has a unique solution (Theorem 2.33(i)), but is strongly singular.

$(\mathcal{D}_A^+)$ is weakly singular, but is not uniquely solvable if $\det \mathcal{C} = 0$ (Theorems 2.38 and 2.39).

$(\mathcal{D}_M^+)$ is weakly singular and has a unique solution (Theorem 2.42), but the computation of this solution requires prior knowledge of the characteristic matrix $\mathcal{C}$, which is not readily available.

$(\mathcal{D}_R^+)$ is weakly singular and has a unique solution that can be computed exclusively from the data (Theorem 2.45).

$(\mathcal{D}_D^+)$ is weakly singular and, under the additional "physical" condition $p\psi = 0$, has a unique solution even when $\mathcal{C}$ is singular (Theorem 2.49(i)).

$(\mathcal{D}_S^+)$ has a unique solution (Theorem 2.52(i)), but is strongly singular.

Consequently, if we consider only the interior Dirichlet problem, then the two clear favourites are the refined indirect method and the direct method. However, the picture changes if we widen the discussion to include the exterior Dirichlet problem. The solution of $(\mathcal{D}_D^-)$ makes explicit use of the matrix Φ (Theorem 2.49(ii)), which is not known a priori. On the other hand, the refined indirect method can solve the more general exterior Dirichlet problem (D_G^-) (see §2.9), and it does so by means of the same equation $(\mathcal{D}_R^+)$ with a simple change of data. As an added bonus, this technique also enables us to calculate the matrices $\mathcal{C}$ and Φ (Remark 2.48(ii)).

It is therefore clear that the refined indirect method offers the best overall computational prospects.

Appendix

A1. Geometry of the boundary curve

For simplicity, we use the same symbol to indicate both a point and its position vector in $\mathbb{R}^2$. Also, c with or without subscripts denotes various positive constants.

Suppose that the boundary ∂S is given in terms of its arc coordinate s by an equation of the form

$$x = \psi(s), \quad s \in [0, l], \quad \psi(0) = \psi(l),$$

with the inverse relationship written as $s = s(x)$, $x \in \partial S$; here l is the length of ∂S. Since ∂S is a C^2-curve, we also have

$$\partial_s \psi(0+) = \partial_s \psi(l-),$$
$$\partial_s^2 \psi(0+) = \partial_s^2 \psi(l-),$$

where $\partial_s = \partial/\partial s$.

Denoting by $\tau(x)$ the unit tangent to ∂S at $x \in \partial S$, orientated in the direction in which s increases, which is chosen so that $\{\nu(x), \tau(x)\}$ is right-handed, we can write

$$\tau_\alpha = \varepsilon_{\beta\alpha} \nu_\beta, \tag{A1}$$

where $\varepsilon_{\alpha\beta}$ is the the two-dimensional Ricci tensor. If $\kappa(x)$ is the algebraic value of the curvature at $x \in \partial S$, then

$$\begin{aligned}\partial_s \tau(x) &= -\kappa(x) \nu(x),\\ \partial_s \nu(x) &= \kappa(x) \tau(x).\end{aligned} \tag{A2}$$

Let

$$\langle x, y \rangle = x_\alpha y_\alpha, \quad |x|^2 = x_1^2 + x_2^2$$

be the standard inner product and the Euclidean norm in $\mathbb{R}^2$.

A1. Lemma. *There is a constant $q > 0$ such that*

$$|\langle \nu(x), x - y \rangle| \leq q|x - y|^2, \tag{A3}$$
$$|\nu(x) - \nu(y)| \leq q|x - y| \tag{A4}$$

for all $x, y \in \partial S$.

Proof. Since ∂S is a C^2-curve, we can define

$$\kappa_0 = \sup_{x \in \partial S} |\kappa(x)|. \tag{A5}$$

It is obvious that $\kappa_0 > 0$, for $\kappa_0 = 0$ would imply that ∂S was a straight line, therefore, not a closed curve.

Let $x = \psi(s)$ and $y = \psi(t)$. By (A2),

$$\partial_{s(x)} |x-y|^2 = 2\langle \tau(x), x-y \rangle,$$
$$\partial^2_{s(x)} |x-y|^2 = 2\bigl[1 - \kappa(x)\langle \nu(x), x-y \rangle\bigr],$$

so for $y \in \partial S$ sufficiently close to x

$$|x-y|^2 = \bigl[1 - \kappa(x')\langle \nu(x'), x'-y \rangle\bigr](s-t)^2,$$
$$\langle \nu(y), x-y \rangle = -\tfrac{1}{2}\kappa(x'')\langle \nu(x''), \nu(y) \rangle (s-t)^2,$$

where $x', x'' \in \partial S$ lie between x and y.

Suppose first that $|x-y| \leq (2\kappa_0)^{-1}$. Then $|x-y|^2 \geq \tfrac{1}{2}(s-t)^2$; consequently,

$$|\langle \nu(y), x-y \rangle| \leq \kappa_0 |x-y|^2.$$

For $|x-y| > (2\kappa_0)^{-1}$ we have $|\langle \nu(y), x-y \rangle| \leq |x-y| \leq l < 4\kappa_0^2 l\,|x-y|^2$; hence, for all $x, y \in \partial S$

$$|\langle \nu(y), x-y \rangle| \leq \max\{\kappa_0, 4\kappa_0^2\}|x-y|^2.$$

Next, $\nu_\alpha(x) - \nu_\alpha(y) = \kappa(x''')\tau_\alpha(x''')(s-t)$, where $x''' \in \partial S$ lies between x and y. Reasoning as above, we find that for all $x, y \in \partial S$

$$|\nu(x) - \nu(y)| \leq 8\kappa_0 |x-y|.$$

The inequalities (A3) and (A4) are now obtained by setting

$$q = \max\{8\kappa_0, 4\kappa_0^2 l\}.$$

A2. Lemma. *Let $x, y \in \partial S$, and let α be the angle between $\nu(x)$ and $\nu(y)$, and γ the acute angle between the support lines of $\nu(x)$ and $x-y$. If $0 < r = \text{const} \leq (2q)^{-1}$, where q is the constant specified in Lemma A1, then for all x and y such that $|x-y| \leq r$ we have*

(i) $\tfrac{1}{2} \leq \cos\alpha \leq 1$;

(ii) $\tfrac{1}{2} \leq \sin\gamma \leq 1$.

Proof. (i) By Lemma A1,

$$\langle \nu(x), \nu(y) \rangle = 1 - \langle \nu(x), \nu(x) - \nu(y) \rangle \geq 1 - qr \geq \tfrac{1}{2}.$$

(ii) By the Mean Value Theorem, there is $x' \in \partial S$ between x and y such that $\tau(x')$ is parallel to $x - y$. By (i), the acute angle β between $\tau(x)$ and $x - y$ satisfies $\tfrac{1}{2} \leq \cos \beta \leq 1$. Statement (ii) now follows from the fact that $\sin \gamma = \cos \beta$.

A3. Lemma. *If*

$$\Sigma_{x,r} = \{y \in \partial S : |x - y| \leq r\}, \quad x \in \partial S, \tag{A6}$$

with r as in Lemma A2, then for all $x \in \partial S$ and $y \in \Sigma_{x,r}$

$$\tfrac{1}{2}|s - t| \leq |x - y| \leq |s - t|, \tag{A7}$$

where $x = \psi(s)$ and $y = \psi(t)$.

Proof. Let

$$a = \psi(s_1), \quad b = \psi(s_2)$$

be the end-points of the arc $\Sigma_{x,r}$. Without loss of generality, we may assume that $0 \leq s_1 < s_2$. For any $y \in \Sigma_{x,r}$ between x and b

$$\partial_t |x - y| = -\frac{\langle \tau(y), x - y \rangle}{|x - y|} = \cos \beta(t),$$

where β is the acute angle between the support lines of $\tau(y)$ and $x - y$; hence,

$$|x - y| = \int_s^t \cos \beta(\sigma) \, d\sigma.$$

By the Mean Value Theorem, there is $\eta \in \partial S$ between x and y such that

$$|x - y| = (t - s) \cos \beta(\eta).$$

Similarly, for any $y \in \Sigma_{x,r}$ between a and x

$$\partial_t |x - y| = -\cos \beta(t),$$

so that, in general, for any $y \in \Sigma_{x,r}$ we can write $|x - y| = c(s, t)|s - t|$, where, by Lemma A2, $\tfrac{1}{2} \leq c(s, t) \leq 1$.

A4. Remark. From the proof of Lemma A3 it is clear that $|x - y|$ is a monotonic function of t on the intervals $I_1 = \{t : y \in \Sigma_{x,r}, t \leq s\}$ and $I_2 = \{t : y \in \Sigma_{x,r}, t \geq s\}$, decreasing on the former and increasing on the latter. This obviously implies that $|x - y'| \neq |x - y''|$ for all $y', y'' \in \Sigma_{x,r}$ such that $y' = \psi(t')$, $y'' = \psi(t'')$, $t' \neq t''$, $t', t'' \in I_1$ or $t', t'' \in I_2$, and that there is a bijective correspondence between the points of the arc $\Sigma_{x,r}$ and those of its projection on the tangent to ∂S at x.

A5. Remark. A slightly modified pair of inequalities (A7) holds for all $x, y \in \partial S$ if by $|s-t|$ we understand the length of the shorter arc of ∂S joining x and y. Since $|s-t| \leq l \leq lr^{-1}|x-y|$ for $|x-y| > r$, we conclude that for all $x, y \in \partial S$

$$c|s-t| \leq |x-y| \leq |s-t|,$$

where $c = \min\{\tfrac{1}{2}, rl^{-1}\}$.

A2. Properties of the boundary layer

Many of the results in this Appendix are proved by considering the behaviour of certain two-point functions in the neighbourhood of the boundary. To help the fluency of such proofs, here we make a preliminary examination of some frequently used properties.

A6. Lemma. *The curves*

$$\partial S_\sigma = \{x \in \mathbb{R}^2 : x = \xi + \sigma\nu(\xi), \xi \in \partial S\}, \quad \sigma = \text{const}, \quad 0 < |\sigma| < \kappa_0^{-1},$$

where κ_0 is given by (A5), are parallel to ∂S and well defined; that is, the support lines of $\nu(\xi)$ and $\nu(\xi')$ do not intersect between $\partial S_{-1/\kappa_0}$ and $\partial S_{1/\kappa_0}$ for any points $\xi, \xi' \in \partial S$.

Proof. Let ϖ be the arc coordinate on ∂S_σ. Then

$$\tau(x) = \frac{dx}{d\varpi} = \left(\frac{d\xi}{ds} + \sigma\frac{d\nu}{ds}\right)\frac{ds}{d\varpi} = [1 + \sigma\kappa(\xi)]\tau(\xi)\frac{ds}{d\varpi}.$$

Since $(d\varpi)^2 = |dx|^2 = [1 + \sigma\kappa(\xi)]^2(ds)^2$ and $1 + \sigma\kappa(\xi) \geq 1 - |\sigma|\kappa_0 > 0$, it follows that $\tau(x) = \tau(\xi)$.

If for some x we have

$$x = \xi + \sigma\nu(\xi) = \xi' + \sigma'\nu(\xi'), \quad |\sigma|, |\sigma'| < \kappa_0^{-1}, \quad \xi \neq \xi',$$

then $\tau(x) = \tau(\xi) = \tau(\xi')$. This leads to $\nu(\xi) = \nu(\xi')$, which contradicts the above assumption, so the assertion is proved.

A7. Lemma. *Consider the boundary layer*

$$S_{\sigma_0} = \{x \in \mathbb{R}^2 : x = \xi + \sigma\nu(\xi),\ \xi \in \partial S,\ |\sigma| \leq \sigma_0\}.$$

If

$$x, x' \in S_{r/4}, \quad |x - x'| < \frac{1}{4}r, \quad x' = \xi' + \sigma'\nu(\xi'), \quad r \leq \min\{\kappa_0^{-1}, (2q)^{-1}\},$$

where κ_0 is defined by (A5) and q is the number specified in Lemma A1, then

$$|\xi - \xi'| < 4|x - x'|.$$

Proof. Without loss of generality, suppose that $|x - \xi| \geq |x' - \xi'|$. Let ξ_0 be the point of intersection of the support lines of $\nu(\xi)$ and $\nu(\xi')$, and η the point on the line through ξ and ξ_0 such that $\eta - x'$ is parallel to $\xi - \xi'$. By Lemma A6, since ξ_0 does not lie between $\partial S_{-1/\kappa_0}$ and $\partial S_{1/\kappa_0}$, we have $|\xi_0 - \xi| > (2\kappa_0)^{-1}$; therefore,

$$\frac{|\eta - x'|}{|\xi - \xi'|} = 1 - \frac{|\eta - \xi|}{|\xi_0 - \xi|} > 1 - \frac{r/4}{1/(2\kappa_0)} = 1 - \tfrac{1}{2}\kappa_0 r \geq \tfrac{1}{2}. \tag{A8}$$

Let ϑ be the angle between $\xi - \xi_0$ and $x' - x$, and γ the acute angle between $\nu(\xi)$ and $\xi' - \xi$. By (A8),

$$|\xi - \xi'| \leq 2|\eta - x'| \leq 2(|x - x'| + |\eta - x|) < 2\bigl(\tfrac{1}{4}r + \tfrac{1}{4}r\bigr) = r,$$

so, by Lemma A2 and (A8),

$$|x - x'| = \frac{\sin\gamma}{\sin\vartheta}|\eta - x'| \geq \tfrac{1}{2}|\eta - x'| > \tfrac{1}{4}|\xi - \xi'|,$$

as required.

A8. Lemma. *With the notation in Lemma A7, if $x, x' \in S_{r/4}$ satisfy*

$$|x - x'| < \frac{1}{8}r, \quad r \leq \min\{\kappa_0^{-1}, (2q)^{-1}\}, \quad \xi = \psi(s),\ \xi' = \psi(s'),\ y = \psi(t),$$

and

$$\Sigma_1 = \{y \in \Sigma_{\xi,r} : |s - t| \leq 8|x - x'|\}, \tag{A9}$$

where $\Sigma_{\xi,r}$ is defined by (A6), then $\xi' \in \Sigma_1$, and for all $y \in \Sigma_1$

(i) $|x - y| \geq \tfrac{1}{2}|\xi - y|$;

(ii) $|x - y| \geq \tfrac{1}{2}|x - \xi|$;

(iii) $|x' - y| \geq \tfrac{1}{2}|\xi' - y| \geq \tfrac{1}{4}|s' - t|$.

Proof. By Lemma A7, $|\xi - \xi'| < 4|x - x'| < r$, so, by Lemmas A3 and A7,

$$|s - s'| \leq 2|\xi - \xi'| < 8|x - x'|,$$

which implies that $\xi' \in \Sigma_1$.

(i) Let γ be the acute angle between $\nu(\xi)$ and $\xi - y$, and ϑ the angle between $x - \xi$ and $x - y$. Since $|\xi - y| \leq r$, Lemma A2 yields

$$|x - y| = \frac{\sin \gamma}{\sin \vartheta}|\xi - y| \geq \tfrac{1}{2}|\xi - y|.$$

(ii) As above,

$$|x - y| = \frac{\sin \gamma}{\sin(\gamma + \vartheta)}|x - \xi| \geq \tfrac{1}{2}|x - \xi|.$$

(iii) The fact that $\xi' \in \Sigma_1$ implies that $|\xi' - y| \leq |s' - t| \leq 8|x - x'| < r$. Repeating the argument in (i) with ξ' instead of ξ, we obtain $|x' - y| \geq \tfrac{1}{2}|\xi' - y|$. Also, by Lemma 1.3, $|\xi' - y| \geq \tfrac{1}{2}|s' - t|$, which completes the proof.

A9. Lemma. *With the notation in Lemma A8, if $x, x' \in S_{r/4}$ satisfy*

$$|x - x'| < \frac{1}{8}r, \quad r \leq \min\{\kappa_0^{-1}, (2q)^{-1}\}$$

and

$$\Sigma_2 = \Sigma_{\xi,r} \setminus \Sigma_1 = \{y \in \Sigma_{\xi,r} : |s - t| > 8|x - x'|\}, \tag{A10}$$

then $\operatorname{mes} \Sigma_2 > 0$, and for all $y \in \Sigma_2$

(i) $|x - y| \geq \tfrac{1}{2}|\xi - y|$;
(ii) $|x' - y| \geq \tfrac{1}{4}|\xi - y|$;
(iii) $|x - x'| < \tfrac{1}{2}|x - y|$;
(iv) $|\xi' - y| < 3|\xi - y|$.

Proof. Since $\operatorname{mes} \Sigma_1 \leq 16|x - x'| < 2r \leq \operatorname{mes} \Sigma$, it follows immediately that $\operatorname{mes} \Sigma_2 > 0$.

(i) This is proved exactly as the first assertion in Lemma A8.

(ii) By Lemma A3 and (i),

$$|x' - y| \geq |x - y| - |x - x'| \geq |x - y| - \tfrac{1}{8}|s - t| \geq \tfrac{1}{2}|\xi - y| - \tfrac{1}{4}|\xi - y| = \tfrac{1}{4}|\xi - y|.$$

(iii) By Lemma A7, $|\xi - \xi'| < 4|x - x'| < r$. Applying Lemma A3 and (i), we now obtain

$$|x - x'| < \tfrac{1}{8}|s - t| \leq \tfrac{1}{4}|\xi - y| \leq \tfrac{1}{2}|x - y|.$$

(iv) By Lemma A8, $\xi' \in \Sigma_1$; consequently, using Lemma A3, we find that

$$|\xi' - y| \leq |\xi - y| + |\xi - \xi'| \leq |\xi - y| + |s - s'| < |\xi - y| + |s - t| \leq 3|\xi - y|,$$

as required.

A10. Lemma. *With the notation in Lemma A8, if the points x, $x' \in S_{r/4}$ satisfy $|x - x'| < \frac{1}{4}r$ and $r \leq \min\{\kappa_0^{-1}, (2q)^{-1}\}$, and $\Sigma_{\xi,r}$ is defined by (A6), then for all $y \in \partial S \setminus \Sigma_{\xi,r}$*

(i) $|x - x'| < \frac{1}{2}|x - y|$;

(ii) $|x - y| > \frac{3}{4}|\xi - y|$;

(iii) $|x' - y| > \frac{1}{4}|\xi - y|$;

(iv) $|\xi' - y| < 2|\xi - y|$.

Proof. Let $y \in \partial S \setminus \Sigma_{\xi,r}$.

(i) Since $x \in S_{r/4}$,

$$|x - y| \geq |\xi - y| - |x - \xi| > r - \tfrac{1}{4}r = \tfrac{3}{4}r > 2|x - x'|.$$

(ii) As above,

$$|x - y| \geq |\xi - y| - \tfrac{1}{4}r > |\xi - y| - \tfrac{1}{4}|\xi - y| = \tfrac{3}{4}|\xi - y|.$$

(iii) By (ii),

$$|x' - y| \geq |x - y| - |x - x'| > |x - y| - \tfrac{1}{2}r > \tfrac{3}{4}|\xi - y| - \tfrac{1}{2}|\xi - y| = \tfrac{1}{4}|\xi - y|.$$

(iv) By Lemma A7,

$$|\xi' - y| \leq |\xi - y| + |\xi - \xi'| < |\xi - y| + 4|x - x'| < |\xi - y| + r < 2|\xi - y|.$$

A11. Lemma. *With the notation in Lemmas A8 and A9, if x, $x' \in S_{r/4}$ satisfy $0 < |x - x'| < \frac{1}{8}r$, $r \leq \min\{\frac{1}{2}, \kappa_0^{-1}, (2q)^{-1}\}$, then there are positive constants c_1, c_2, c_3 and c_4 such that*

(i) $\int_{\Sigma_1} |x - y|^{-\gamma} ds(y) \leq c_1 |x - x'|^{1-\gamma}$ *for any $\gamma < 1$, where c_1 depends on γ;*

(ii) $\int_{\Sigma_1} |x' - y|^{-\gamma} ds(y) \leq c_2 |x - x'|^{1-\gamma}$ *for any $\gamma < 1$, where c_2 depends on γ;*

(iii) $\int_{\Sigma_2} |x - y|^{-\gamma-1} ds(y) \leq c_3 |x - x'|^{-\gamma}$ *for any $\gamma \in (0,1)$, where c_3 depends on γ;*

(iv) $\int_{\Sigma_2} |x - y|^{-1} ds(y) \leq c_4 |\ln |x - x'||$.

Proof. Let $\delta = |x - x'|$, $x = \xi + \sigma\nu(\xi)$, $x' = \xi' + \sigma'\nu(\xi')$, $\xi, \xi' \in \partial S$, $\xi = \psi(s)$, $\xi' = \psi(s')$, $y = \psi(t)$, and

$$\Gamma_1 = \{t : y \in \Sigma_1\} = \{t : |s - t| \le 8\delta\},$$
$$\Gamma_2 = \{t : y \in \Sigma_2\} = \{t : y \in \Sigma_{\xi,r}, |s - t| > 8\delta\}. \tag{A11}$$

Without loss of generality, suppose that the point corresponding to the origin of the arc coordinate lies outside $\Sigma_{\xi,r}$.

(i) By Lemmas A8 and A3,

$$\int_{\Sigma_1} |x - y|^{-\gamma} ds(y) \le 2^\gamma \int_{\Sigma_1} |\xi - y|^{-\gamma} ds(y) \le 4^\gamma \int_{\Gamma_1} |s - t|^{-\gamma} dt = \frac{2 \cdot 4^\gamma \cdot 8^{1-\gamma}}{1 - \gamma} \delta^{1-\gamma}.$$

(ii) By Lemma A8 and the definition of Γ_1 in (A11), $s' \in \Gamma_1$, which means that $|s - s'| \le 8\delta$; therefore, as above,

$$\int_{\Sigma_1} |x' - y|^{-\gamma} ds(y) \le 4^\gamma \int_{\Gamma_1} |s' - t|^{-\gamma} dt$$
$$\le \frac{2 \cdot 4^\gamma}{1 - \gamma}(|s - s'| + 8\delta)^{1-\gamma} \le \frac{2 \cdot 4^\gamma \cdot 16^{1-\gamma}}{1 - \gamma} \delta^{1-\gamma}.$$

(iii) Applying Lemmas A9 and A3 and calculating the integral explicitly in terms of the end-points of Γ_2, we obtain

$$\int_{\Sigma_2} |x - y|^{-\gamma-1} ds(y) \le 4^{\gamma+1} \int_{\Gamma_2} |s - t|^{-\gamma-1} dt \le \frac{8}{\gamma} \delta^{-\gamma}.$$

(iv) If a and b are the end-points of $\Sigma_{\xi,r}$, $a = \psi(s_a)$, $b = \psi(s_b)$, $s_a < s < s_b$, then

$$\int_{\Sigma_2} |x - y|^{-1} ds(y) \le 4 \int_{\Gamma_2} |s - t|^{-1} dt = 4\big[\ln(s - s_a) + \ln(s_b - s) - 2\ln(8\delta)\big].$$

Since $r < \frac{1}{2}$, from Lemma A3 it follows that $s - s_a \le 2|x - a| = 2r < 1$ and, similarly, $s_b - s < 1$; hence,

$$\int_{\Sigma_2} |x - y|^{-1} ds(y) < 8|\ln \delta|,$$

which completes the proof.

A12. Remark. It is obvious that all the conditions in Lemmas A2, A3, A6, A10 and A11 are satisfied if, for example, we choose x, $x' \in S_{r/4}$ such that

$$|x - x'| < \tfrac{1}{8}r, \quad r = \min\{\tfrac{1}{2}, \kappa_0^{-1}, (2q)^{-1}\}. \tag{A12}$$

From now on we use the notation

$$S_0 = S_{r/4}, \quad S_0^+ = \{x \in S_0 : x = \xi + \sigma\nu(\xi),\ \xi \in \partial S,\ -\tfrac{1}{4}r \leq \sigma < 0\}, \quad S_0^- = S_0 \setminus \bar{S}_0^+.$$

A13. Remark. For $x \in \partial S$ we introduce local coordinates (ρ, ω) along the positive tangent and inward normal to ∂S at x, respectively. Since ∂S is a simple C^2-curve, in accordance with Remark A4 there is a twice continuously differentiable function f on $[-r, r]$ satisfying $f(0) = f'(0) = 0$ and such that the equation of $\Sigma_{x,r}$ can be written in the form $\omega = f(\rho)$. If α is the angle between $\nu(x)$ and $\nu(y)$ and β is the acute angle between the support lines of $x - y$ and $\tau(x)$, then, by Lemma A2, we find that for all $y \in \Sigma_{x,r}$

$$|f(\rho)| = |x - y|\sin\beta \leq \tfrac{1}{2}\sqrt{3}\,r,$$
$$|f'(\rho)| = |\tan\alpha| \leq \sqrt{3}, \quad |f''(\rho)| \leq \kappa_0\bigl[1 + f'^2(\rho)\bigr]^{3/2} \leq \kappa_0.$$

A14. Theorem. *If x, $y \in \partial S$, $x = \psi(s)$ and $y = \psi(t)$, then*

$$\frac{|x - y|}{|s - t|} \to 1 \quad \text{as } y \to x,$$

uniformly on ∂S.

Proof. Using the local coordinates (ρ, ω) defined in Remark A13, we can write

$$f(\rho) = \tfrac{1}{2}\rho^2 f''(\rho_1), \quad f'(\rho) = \rho f''(\rho_2), \tag{A13}$$

with ρ_1 and ρ_2 between 0 and ρ. Consequently, for $\rho > 0$

$$\begin{aligned} f_1(\rho) &= |x - y| = \bigl[\rho^2 + f^2(\rho)\bigr]^{1/2} = \rho + \tfrac{1}{2}\rho^2 f_1''(\rho'), \\ f_2(\rho) &= |s - t| = \int_0^\rho \bigl[1 + f'^2(\theta)\bigr]^{1/2}\,d\theta = \rho + \tfrac{1}{2}\rho^2 f_2''(\rho''), \end{aligned} \tag{A14}$$

where $0 < \rho'$, $\rho'' < \rho$. From this we find that

$$\left|\frac{f_1(\rho)}{f_2(\rho)} - 1\right| = \frac{\rho|f_1''(\rho') - f_2''(\rho'')|}{|2 + \rho f_2''(\rho'')|}.$$

Now (A13) and (A14) yield

$$f_1''(\rho) = \rho\left[\tfrac{1}{4}f''^2(\rho_1) + f''^2(\rho_2) + \tfrac{1}{2}f''(\rho_1)f''(\rho) - f''(\rho_1)f''(\rho_2)\right.$$
$$\left. + \tfrac{1}{8}\rho^2 f''^3(\rho_1)f''(\rho)\right]\left[1 + \tfrac{1}{4}\rho^2 f''^2(\rho_1)\right]^{-3/2},$$
$$f_2''(\rho) = f'(\rho)f''(\rho)\left[1 + f'^2(\rho)\right]^{-1/2},$$

so that, by Remark A13,

$$|f_1''(\rho)| \leq c_1, \quad |f_2''(\rho)| \leq c_1, \quad c_1 = \text{const} > 0,$$

for all $\rho \in [-r, r]$ and all $x \in \partial S$.

Let $0 < \rho < c_1^{-1}$. Since

$$|2 + \rho f_2''(\rho'')| \geq 2 - \rho|f_2''(\rho'')| > 1,$$

we conclude that

$$\left|\frac{f_1(\rho)}{f_2(\rho)} - 1\right| \leq c_2 \rho,$$

where the positive constant c_2 is independent of x.

The same inequality (with $|\rho|$ on the right-hand side) is obtained for $-c_1^{-1} < \rho < 0$.

A15. Theorem. *With the notation in Theorem A14,*

$$\frac{|x - y|}{\rho} \to 1 \quad \text{as } y \to x,$$

uniformly on ∂S.

Proof. As above, we find that

$$\left|\frac{f_1(\rho)}{\rho} - 1\right| = \frac{\tfrac{1}{4}|\rho|^2 f''^2(\rho_1)}{\left[1 + \tfrac{1}{4}\rho^2 f''^2(\rho_1)\right]^{1/2} + 1} \leq c|\rho|^2,$$

where ρ_1 lies between 0 and ρ and c is independent of x.

A16. Remark. If f is continuously differentiable in S_0, then

$$(\operatorname{grad} f)(x) = \left[\tau(x)\partial_{s(x)} + \nu(x)\partial_{\nu(x)}\right]f(x), \quad x \in \partial S. \tag{A15}$$

A3. Integrals with singular kernels

We write S for either S^+ or S^-, consider the set of all functions in $C(S)$ ($C^1(S)$) that are continuously extendable (continuously extendable together with their first order derivatives) to $\bar{S} = S \cup \partial S$, and denote by $C(\bar{S})$ ($C^1(\bar{S})$) the space of the corresponding extensions. The following assertion shows that this notation is justified.

A17. Theorem. *Let $f \in C^1(S)$, and let $f(x) \to p(\xi)$ and $(\operatorname{grad} f)(x) \to q(\xi)$ as $S \ni x \to \xi \in \partial S$, where p and q are continuous on ∂S. Then the function*

$$\tilde{f}(x) = \begin{cases} f(x), & x \in S, \\ p(x), & x \in \partial S, \end{cases}$$

has (one-sided) derivatives at all $x \in \partial S$ and

$$(\operatorname{grad} \tilde{f})(x) = \begin{cases} (\operatorname{grad} f)(x), & x \in S, \\ q(x), & x \in \partial S \end{cases}$$

(that is, the operations of differentiation and extension to $\bar{S}$ commute for f).

Proof. Clearly, $\tilde{f} \in C(\bar{S}) \cap C^1(S)$. Consequently, for points $\xi = (\xi_1, \xi_2) \in \partial S$ and $x = (x_1, \xi_2) \in S$, $x_1 \neq \xi_1$, in a sufficiently small neighbourhood of ξ we have

$$\left| \frac{\tilde{f}(x) - \tilde{f}(\xi)}{x_1 - \xi_1} - q_1(\xi) \right| = \left| \partial_{x_1} \tilde{f}(\eta) - q_1(\xi) \right| = \left| \partial_{x_1} f(\eta) - q_1(\xi) \right|,$$

where $\eta = (\eta_1, \xi_2)$ with η_1 between x_1 and ξ_1. The result for $\partial_{x_1} \tilde{f}$ now follows from the fact that the right-hand side tends to zero as $x \to \xi$. The argument for $\partial_{x_2} \tilde{f}$ is similar.

A18. Remark. The above spaces are also introduced for functions defined on ∂S. Let $f(x)$ be such a function, and let $x = \psi(s)$ in terms of the arc coordinate. Then for simplicity we also write $f(s) \equiv f(\psi(s))$. In this case, the derivative of f is defined to be

$$f'(s) = f'(x) = \partial_s f(x) = \lim_{t \to s} \frac{f(y) - f(x)}{t - s} = \lim_{t \to s} \frac{f(t) - f(s)}{t - s},$$

where $x, y \in \partial S$ and $y = \psi(t)$, provided that the limit exists. We specify that in what follows the notation f' for the derivative does not extend to position vectors. Thus, x' will denote a point on ∂S and not dx/ds.

Clearly, if f is defined and differentiable on a domain that includes ∂S, then the derivative along ∂S of the restriction of f to ∂S coincides with $\langle (\operatorname{grad} f)(x), \tau(x) \rangle$.

A19. Lemma. *If $0 < \beta < \alpha \leq 1$, then*

(i) $C^{0,\alpha}(\bar{S}) \subset C^{0,\beta}(\bar{S})$;

(ii) $fg \in C^{0,\beta}(\bar{S})$ for all $f \in C^{0,\alpha}(\bar{S})$ and $g \in C^{0,\beta}(\bar{S})$.

The proof consists in the verification of the required properties.

In view of Lemma A3, we do not distinguish between $C^{0,\alpha}(\partial S)$ and $C^{0,\alpha}([0,l])$, which is defined by means of the inequality

$$|f(s) - f(t)| \leq c|s - t|^\alpha \quad \text{for all } s, t \in [0, l].$$

Obviously, Lemma A19 also holds for functions on ∂S.

A20. Remark. If f is bounded in $\bar{S}$, that is, $|f(x)| \leq M = \text{const}$ for all $x \in \bar{S}$, and (1.1) holds for all $x, y \in \bar{S}$ such that $|x - y| \leq \delta$, where $\delta = \text{const} > 0$, then it holds (possibly with a different c) for all $x, y \in \bar{S}$. This is easily seen, since for $|x - y| > \delta$ we can write

$$|f(x) - f(y)| \leq 2M < 2M\delta^{-\alpha}|x - y|^\alpha.$$

A21. Remark. If $\varphi \in C^{0,\alpha}(\partial S)$ as a function of $x = \psi(s)$, then, by Lemma A3, $\varphi \in C^{0,\alpha}(\partial S)$ also as a function of s, and vice versa.

A22. Remark. A kernel $k(x, y)$ may have a lower "singularity index" γ (see Definition 1.2) when it is considered on ∂S rather than in S_0. For example, the function $k(x, y) = \partial_{\nu(y)} \ln |x - y|$ is a proper 1-singular kernel in S_0, but, by Lemma A1, a proper 0-singular kernel on ∂S.

A23. Lemma. *If $k(x,y)$ is γ-singular in S_0, $\gamma \in [0,1]$, and continuously differentiable with respect to x_α for all $x \in S_0$ and $y \in \partial S$, $x \neq y$, and if the kernels $|x - y|\partial_{x_\alpha} k(x, y)$ are γ-singular in S_0, then $k(x,y)$ is a proper γ-singular kernel in S_0.*

Proof. Let $x, x' \in S_0$ and $y \in \partial S$ be such that $0 < |x - x'| < \frac{1}{2}|x - y|$. For any x'' on the line between x and x' we have

$$|x'' - y| \geq |x - y| - |x - x''| > |x - y| - \tfrac{1}{2}|x - y| = \tfrac{1}{2}|x - y|;$$

consequently,

$$|k(x, y) - k(x', y)| \leq |x_\alpha - x'_\alpha| |\partial_{x_\alpha} k(x'', y)| \leq c|x - x'|\, |x - y|^{-\gamma - 1},$$

where $c = \text{const}$ depends only on γ.

A24. Remark. If $k(x,y)$ is a γ-singular kernel on ∂S, $\gamma \in [0,1]$, and continuously differentiable with respect to the arc coordinate s of x at all points $x \in \partial S$, $x \neq y$, and if $|x - y|\partial_{s(x)}k(x,y)$ is γ-singular on ∂S, then $k(x,y)$ is a proper γ-singular kernel on ∂S. The proof of this statement is similar to that of Lemma A23, use also being made of Remark A5.

The following assertion is proved by direct verification.

A25. Lemma. (i) If $k_1(x,y)$ is 0-singular and $k_2(x,y)$ is γ-singular, $\gamma \in [0,1]$, then $k_1(x,y)k_2(x,y)$ is γ-singular.

(ii) If $k_1(x,y)$ is γ_1-singular and $k_2(x,y)$ is γ_2-singular, $0 \leq \gamma_1 \leq \gamma_2 \leq 1$, then $k_1(x,y) + k_2(x,y)$ is γ_2-singular.

A26. Remark. Lemma 1.27 also holds with "singular" replaced by "proper singular" in its statement.

A27. Theorem. If $k(x,y)$ is a γ-singular kernel on ∂S, $\gamma \in [0,1)$, then the function

$$f(x) = \int_{\partial S} k(x,y)\, ds(y) \tag{A16}$$

is continuous on ∂S.

Proof. Let $x, a, b, y \in \partial S$, $x = \psi(s)$, $a = \psi(s - \varepsilon_1)$, $b = \psi(s + \varepsilon_2)$, $y = \psi(t)$, where $\varepsilon_1, \varepsilon_2 > 0$ are arbitrarily small, and let

$$I_\varepsilon(s) = \int_b^a |s - t|^{-\gamma}\, dt, \quad I(s) = \int_{\partial S} |s - t|^{-\gamma}\, dt.$$

Clearly,

$$|I(s) - I_\varepsilon(s)| = \frac{1}{1-\gamma}(\varepsilon_1^{1-\gamma} + \varepsilon_2^{1-\gamma});$$

therefore, $I_\varepsilon(s) \to I(s)$ uniformly with respect to s as $\varepsilon_1, \varepsilon_2 \to 0$. Since, by Definition 1.2 and Lemma A3,

$$|k(x,y)| \leq c|x-y|^{-\gamma} \leq c|s-t|^{-\gamma}$$

for all $x, y \in \partial S$, $x \neq y$, the improper integral (A16) converges uniformly with respect to $x \in \partial S$, and the assertion follows from a well-known theorem of analysis (see, for example, [22]).

A28. Theorem. *If $k(x,y)$ is a proper γ-singular kernel in S_0 (on ∂S), $\gamma \in [0,1)$, and $\varphi \in C(\partial S)$, then the function*

$$K(x) = \int_{\partial S} k(x,y)\varphi(y)\, ds(y), \quad x \in S_0 \ (x \in \partial S),$$

belongs to $C^{0,\beta}(S_0)$ ($C^{0,\beta}(\partial S)$), with $\beta = 1 - \gamma$ for $\gamma \in (0,1)$ and any $\beta \in (0,1)$ for $\gamma = 0$. In addition,

$$\sup_{\substack{x,x' \in S_0 \ (\partial S) \\ x \neq x'}} \frac{|K(x) - K(x')|}{|x - x'|^\beta} \leq c \sup_{x \in \partial S} |\varphi(x)|,$$

where $c = \text{const} > 0$ may depend on γ.

Proof. $K(x)$ is obviously an improper integral for $x \in \partial S$.

Let $\Sigma_{x,r}$, Σ_1 and Σ_2 be the sets defined by (A6), (A9) and (A10). In view of Remark A20, we may consider x, $x' \in S_0$ satisfying (A12).

Setting

$$x = \xi + \sigma\nu(\xi), \quad x' = \xi' + \sigma'\nu(\xi'), \quad \xi,\xi' \in \partial S,$$

we can write

$$K(x) - K(x') = I_1 + I_2 + I_3,$$

where, by Definition 1.2, Remark A12 and Lemmas A8–A11,

$$|I_1| = \left| \int_{\Sigma_1} [k(x,y) - k(x',y)]\varphi(y)\, ds(y) \right|$$

$$\leq c_1 \sup_{x \in \partial S} |\varphi(x)| \int_{\Sigma_1} (|x-y|^{-\gamma} + |x'-y|^{-\gamma})\, ds(y)$$

$$\leq c_2 |x - x'|^{1-\gamma} \sup_{x \in \partial S} |\varphi(x)|,$$

$$|I_2| = \left| \int_{\Sigma_2} [k(x,y) - k(x',y)]\varphi(y)\, ds(y) \right|$$

$$\leq c_3 |x - x'| \sup_{x \in \partial S} |\varphi(x)| \int_{\Sigma_2} |x - y|^{-\gamma-1}\, ds(y)$$

$$\leq c_4 |x - x'|^{1-\gamma} \sup_{x \in \partial S} |\varphi(x)| \quad \text{if } \gamma \in (0,1),$$

$$|I_2| \leq c_5 |x - x'| |\ln|x - x'|| \sup_{x \in \partial S} |\varphi(x)| \quad \text{if } \gamma = 0,$$

$$|I_3| = \left| \int_{\partial S \setminus \Sigma_{\xi,r}} [k(x,y) - k(x',y)] \varphi(y) \, ds(y) \right|$$

$$\leq c_6 |x - x'| \sup_{x \in \partial S} |\varphi(x)| \int_{\partial S \setminus \Sigma_{\xi,r}} |x - y|^{-\gamma - 1} \, ds(y)$$

$$\leq c_7 r^{-\gamma - 1} |\partial S| \, |x - x'| \sup_{x \in \partial S} |\varphi(x)| = c_8 |x - x'| \sup_{x \in \partial S} |\varphi(x)|.$$

The assertion now follows from the fact that the constants $c_1, \ldots, c_8 > 0$ are independent of x and x' (although they may depend on γ).

The result is established for $x, x' \in \partial S$ as a particular case of the above, by setting $x = \xi$ and $x' = \xi'$.

A29. Remark. It is obvious that Theorem A28 holds if the kernel $k(x,y)$ is continuous on $S_0 \times \partial S$ ($\partial S \times \partial S$).

A30. Theorem. *If $k(x,y)$ is a proper 1-singular kernel in S_0 (on ∂S), φ belongs to $C^{0,\alpha}(\partial S)$, $\alpha \in (0,1]$, and*

$$\Phi(x) = \int_{\partial S} k(x,y) [\varphi(y) - \varphi(\xi)] \, ds(y), \tag{A17}$$

where $x = \xi + \sigma\nu(\xi) \in S_0$ ($x = \xi \in \partial S$), then $\Phi \in C^{0,\beta}(S_0)$ ($\Phi \in C^{0,\beta}(\partial S)$) for any $\beta \in (0, \alpha)$. If, in addition, $\alpha \in (0,1)$ and

$$\left| \int_{\partial S \setminus \Sigma_{\xi,\delta}} k(x,y) \, ds(y) \right| \leq c = \text{const} > 0 \tag{A18}$$

for all $x \in S_0$ ($x \in \partial S$) and all $0 < \delta < r$, then $\Phi \in C^{0,\alpha}(S_0)$ ($\Phi \in C^{0,\alpha}(\partial S)$).

Proof. Clearly, Φ exists as an improper integral if $x \in \partial S$, and, by Theorem A27, is continuous on ∂S.

As in the proof of Theorem A28, let $x, x' \in S_0$ be chosen so that (A12) holds. Writing

$$\Phi(x) - \Phi(x') = \int_{\Sigma_1} \{ k(x,y) [\varphi(y) - \varphi(\xi)] - k(x',y) [\varphi(y) - \varphi(\xi')] \} \, ds(y)$$

$$+ \int_{\Sigma_2} \{ [k(x,y) - k(x',y)] [\varphi(y) - \varphi(\xi')]$$

$$- k(x,y) [\varphi(\xi) - \varphi(\xi')] \} \, ds(y)$$

$$+ \int_{\partial S \setminus \Sigma_{\xi,r}} \left\{ \left[k(x,y) - k(x',y) \right] \left[\varphi(y) - \varphi(\xi') \right] \right.$$
$$\left. - k(x,y) \left[\varphi(\xi) - \varphi(\xi') \right] \right\} ds(y)$$
$$= I_1 + I_2 + I_3,$$

from Definition 1.1, Remark A12 and Lemma A11 we now find that

$$|I_1| \leq c_1 \int_{\Sigma_1} \left(|\xi - y|^{\alpha-1} + |\xi' - y|^{\alpha-1} \right) ds(y) \leq c_2 |x - x'|^{\alpha},$$

$$|I_2| \leq c_3 |x - x'| \int_{\Sigma_2} |\xi - y|^{\alpha-2} ds(y) + c_4 |\xi - \xi'|^{\alpha} \int_{\Sigma_2} |\xi - y|^{-1} ds(y)$$
$$\leq c_5 |x - x'|^{\alpha} + c_6 |x - x'|^{\alpha} |\ln |x - x'|| \quad \text{for } \alpha \in (0,1),$$

$$|I_2| \leq c_3 |x - x'| \int_{\Sigma_2} |\xi - y|^{-1} ds(y) + c_4 |\xi - \xi'|^{\alpha} \int_{\Sigma_2} |\xi - y|^{-1} ds(y)$$
$$\leq c_7 |x - x'|^{\alpha} |\ln |x - x'|| \quad \text{for } \alpha = 1,$$

$$|I_3| \leq c_8 |x - x'| \int_{\partial S \setminus \Sigma_{\xi,r}} |\xi - y|^{\alpha-2} ds(y) + c_9 |\xi - \xi'|^{\alpha} \int_{\partial S \setminus \Sigma_{\xi,r}} |\xi - y|^{-1} ds(y)$$
$$\leq c_8 r^{\alpha-2} |\partial S| \, |x - x'| + c_{10} r^{-1} |\partial S| \, |x - x'|^{\alpha} \leq c_{11} |x - x'|^{\alpha},$$

where $c_1, \ldots, c_{11}$ are positive constants independent of x and x'.

This proves the first part of the assertion. For the second part we combine the last terms in I_2 and I_3 and use the fact that $\int_{\partial S \setminus \Sigma_1} k(x,y) \, ds(y)$ is bounded for all $x, x' \in S_0$ satisfying the conditions of the theorem. (See Remark A36 below for a full explanation of this detail.)

The result for $x, x' \in \partial S$ is again obtained by setting $x = \xi$ and $x' = \xi'$.

A31. Remark. By Theorem A27, estimate (A18) holds on ∂S if $k(x,y)$ is a γ-singular kernel on ∂S, $\gamma \in [0,1)$.

A32. Theorem. Let $k(x,y)$ be a β-singular kernel on ∂S, $\beta \in [0,1)$, such that

(i) $g(x) = \partial_s \left[\int_{\partial S} k(x,y) \, ds(y) \right]$ exists for all $x \in \partial S$, and $g \in C(\partial S)$;

(ii) $\left[k(x',y) - k(x,y) \right] (s' - s)^{-1} = k_0(x,y) + O(|s' - s| \, |x - y|^{-\gamma-2})$ for all points $x, x', y \in \partial S$, $0 < |x - x'| < \frac{1}{2} |x - y|$, where $x = \psi(s)$, $x' = \psi(s')$, and $|x - y| k_0(x,y)$ is a γ-singular kernel on ∂S, $\gamma \in [0,1)$.

If $\varphi \in C^{0,\alpha}(\partial S)$, $\alpha \in (\beta, 1]$, $\alpha > \gamma$, then the function

$$F(x) = \int_{\partial S} k(x,y)\varphi(y)\,ds(y), \quad x \in \partial S,$$

belongs to $C^1(\partial S)$ and

$$\partial_s F(x) = \int_{\partial S} k_0(x,y)\bigl[\varphi(y) - \varphi(x)\bigr]ds(y) + \varphi(x)g(x). \tag{A19}$$

Proof. Let $G(x)$ be the function on the right-hand side in (A19). By Theorem A27, $F(x)$ and the first term in $G(x)$ exist as improper integrals and are continuous on ∂S; the second term in $G(x)$ is continuous by assumption.

Let $x, x' \in \partial S$ be such that $0 < |x - x'| < \tfrac{1}{8}r$, with r satisfying (A12). We have

$$\bigl[F(x') - F(x)\bigr](s' - s)^{-1} - G(x)$$
$$= (s'-s)^{-1}\int_{\Sigma_1}\bigl\{k(x',y)\bigl[\varphi(y) - \varphi(x')\bigr] - k(x,y)\bigl[\varphi(y) - \varphi(x)\bigr]\bigr\}ds(y)$$
$$+ (s'-s)^{-1}\bigl[\varphi(x') - \varphi(x)\bigr]\int_{\Sigma_1} k(x',y)\,ds(y) - \int_{\Sigma_1} k_0(x,y)\bigl[\varphi(y) - \varphi(x)\bigr]ds(y)$$
$$+ \int_{\Sigma_2}\bigl\{\bigl[k(x',y) - k(x,y)\bigr](s'-s)^{-1} - k_0(x,y)\bigr\}\bigl[\varphi(y) - \varphi(x)\bigr]ds(y)$$
$$+ \int_{\partial S \setminus \Sigma_{x,r}}\bigl\{\bigl[k(x',y) - k(x,y)\bigr](s'-s)^{-1} - k_0(x,y)\bigr\}\bigl[\varphi(y) - \varphi(x)\bigr]ds(y)$$
$$+ \varphi(x)\biggl\{\biggl[\int_{\partial S} k(x',y)\,ds(y) - \int_{\partial S} k(x,y)\,ds(y)\biggr](s'-s)^{-1} - g(x)\biggr\}$$
$$= I_1 + I_2 + I_3 + I_4 + I_5 + I_6.$$

By Definition 1.1, Remark A12 and Lemmas A8 and A11,

$$|I_1| \leq c_1|s'-s|^{-1}\int_{\Sigma_1}\bigl(|x'-y|^{\alpha-\beta} + |x-y|^{\alpha-\beta}\bigr)ds(y) \leq c_2|s'-s|^{\alpha-\beta},$$

$$|I_2| \leq c_3|s'-s|^{\alpha-1}\int_{\Sigma_1}|x'-y|^{-\beta}\,ds(y) \leq c_4|s'-s|^{\alpha-\beta},$$

$$|I_3| \leq c_5\int_{\Sigma_1}|x-y|^{\alpha-\gamma-1}\,ds(y) \leq c_6|s'-s|^{\alpha-\gamma}.$$

By Lemma A9, $y \in \Sigma_2$ implies that $|x - x'| < \frac{1}{2}|x - y|$; hence,

$$|I_4| \leq c_7|s' - s| \int_{\Sigma_2} |x - y|^{\alpha - \gamma - 2} \, ds(y) \leq c_8|s' - s|^{\alpha - \gamma}.$$

Finally, by Lemma A10,

$$|I_5| \leq c_9|s' - s| \int_{\partial S \setminus \Sigma_{x,r}} |x - y|^{\alpha - \gamma - 2} \, ds(y) \leq c_{10}|s' - s|.$$

Since all the constants $c_1, \ldots, c_{10} > 0$ are independent of x and x', we find that $I_j \to 0$ as $s' - s \to 0$, $j = 1, \ldots, 5$.

In addition, by our assumption (i), $I_6 \to 0$ as $s' - s \to 0$, which proves that $F'(x)$ exists for all $x \in \partial S$ and is given by (A19), whose right-hand side is obviously a continuous function on ∂S.

A33. Remark. Under the conditions in Theorem A32, if $g \in C^{0,\alpha}(\partial S)$ and $k_0(x, y)$ is a proper 1-singular kernel on ∂S, then, by Theorem A30, $F \in C^{1,\beta}(\partial S)$ for any $\beta \in (0, \alpha)$. If, furthermore, $\alpha \in (0, 1)$ and $k_0(x, y)$ satisfies the estimate (A18), then $F \in C^{1,\alpha}(\partial S)$.

A34. Remark. In practice it is helpful to have some easily checked condition in place of assumption (ii) in Theorem A32. Suppose that $k(x, y)$ is continuously differentiable with respect to $s(x)$ for all $x, y \in \partial S$, $x \neq y$, and that $|x - y|\partial_{s(x)} k(x, y)$ is a proper γ-singular kernel on ∂S, $\gamma \in [0, 1)$. Then for $x, x', y \in \partial S$ such that $0 < |x - x'| < \frac{1}{2}|x - y|$ we have

$$\big[k(x', y) - k(x, y)\big](s' - s)^{-1} = \partial_{s(x)} k(x, y) + \big[\partial_{s(x)} k(x'', y) - \partial_{s(x)} k(x, y)\big],$$

where $x'' \in \partial S$ lies between x and x'. Since

$$\big|\partial_{s(x)} k(x'', y) - \partial_{s(x)} k(x, y)\big| \leq c|s - s'| \, |x - y|^{-\gamma - 2},$$

it follows that, under the above conditions, assumption (ii) in Theorem A32 holds with $k_0(x, y) = \partial_{s(x)} k(x, y)$.

A35. Definition. Let $k(x, y)$ be defined and continuous for all points $x, y \in \partial S$, $x \neq y$. We say that $\int_{\partial S} k(x, y) \, ds(y)$ exists as *principal value* if

$$\lim_{\delta \to 0} \int_{\partial S \setminus \Sigma_{x,\delta}} k(x, y) \, ds(y) \tag{A20}$$

exists for all $x \in \partial S$.

Obviously, an ordinary (even improper) integral exists as principal value, but the converse is not true in general.

In what follows, the principal value of an integral (if it exists) is denoted by the same symbol as an ordinary integral, the difference in meaning being either explicitly stated, or understood from the context as the only possible alternative.

A36. Remark. Let $k(x,y)$ be a 1-singular kernel on ∂S, and let a_1, a_2 and b_1, b_2, $a_1 \leq b_1 < b_2 \leq a_2$, be the arc coordinates of the end-points of the sets $\Sigma_{x,\delta}$ and

$$\Gamma_{x,\delta} = \{y \in \partial S : |t - s| \leq \delta\}, \tag{A21}$$

respectively, where $x = \psi(s)$ and $y = \psi(t)$. Since

$$\left| \int_{\partial S \setminus \Gamma_{x,\delta}} k(x,y)\, ds(y) - \int_{\partial S \setminus \Sigma_{x,\delta}} k(x,y)\, ds(y) \right|$$

$$= \left| \int_{\Sigma_{x,\delta} \setminus \Gamma_{x,\delta}} k(x,y)\, ds(y) \right| \leq c \left(\int_{a_1}^{b_1} |s - t|^{-1}\, dt + \int_{b_2}^{a_2} |s - t|^{-1}\, dt \right)$$

$$= c \ln\left(\frac{s - a_1}{s - b_1} \cdot \frac{a_2 - s}{b_2 - s} \right) = c \ln\left(\frac{s - a_1}{\delta} \cdot \frac{a_2 - s}{\delta} \right),$$

Theorem A14 implies that if $\int_{\partial S} k(x,y)\, ds(y)$ exists in the sense of principal value, then its definition can equivalently be given as

$$\lim_{\delta \to 0} \int_{\partial S \setminus \Gamma_{x,\delta}} k(x,y)\, ds(y).$$

Moreover, if the limit (A20) exists uniformly for all $x \in \partial S$, then so does the above one, and vice versa.

A37. Remark. Let ρ be the local coordinate of $y \in \Sigma_{x,r}$ measured from x along the support line of $\tau(x)$ (see Remark A13), and consider the set

$$\Lambda_{x,\delta} = \{y \in \Sigma_{x,r} : |\rho| \leq \delta\}, \quad \delta < \tfrac{1}{2}r.$$

Since $\delta < \tfrac{1}{2}r$, all points in the neighbourhood of x such that $|\rho| \leq \delta$ belong to $\Sigma_{x,r}$. Denoting by $-a$ and b, $a, b > 0$, the ρ-coordinates of the end-points of $\Sigma_{x,\delta}$, we find that for a 1-singular kernel $k(x,y)$ on ∂S

$$\left| \int_{\partial S \setminus \Sigma_{x,\delta}} k(x,y)\, ds(y) - \int_{\partial S \setminus \Lambda_{x,\delta}} k(x,y)\, ds(y) \right|$$

$$= \left| \int_{\Lambda_{x,\delta} \setminus \Sigma_{x,\delta}} k(x,y)\, ds(y) \right| \leq c_1 \int_{\Lambda_{x,\delta} \setminus \Sigma_{x,\delta}} |x-y|^{-1}\, ds(y)$$

$$\leq c_2 \left(\int_{-\delta}^{-a} (-\rho)^{-1}\, d\rho + \int_{b}^{\delta} \rho^{-1}\, d\rho \right) = c_2 \ln\left(\frac{\delta}{a} \cdot \frac{\delta}{b} \right),$$

where c_2 does not depend on x. Consequently, by Theorem A15, if $\int_{\partial S} k(x,y)\, ds(y)$ exists in the sense of principal value, then it can also be defined as

$$\lim_{\delta \to 0} \int_{\partial S \setminus \Lambda_{x,\delta}} k(x,y)\, ds(y).$$

Furthermore, from Theorem A15 it follows that the existence of either of these two equivalent limits uniformly with respect to $x \in \partial S$ implies the same property for the other one.

A38. Theorem. *Let $k(x,y)$ be a proper 1-singular kernel in S_0 that is γ-singular on ∂S, $\gamma \in [0,1)$, and let*

$$f(x) = \int_{\partial S} k(x,y)\, ds(y), \quad x \in S_0 \setminus \partial S,$$
$$f_0(x) = \int_{\partial S} k(x,y)\, ds(y), \quad x \in \partial S, \tag{A22}$$

and

$$F(x) = \int_{\partial S} k(x,y)\varphi(y)\, ds(y), \quad x \in S_0 \setminus \partial S,$$
$$F_0(x) = \int_{\partial S} k(x,y)\varphi(y)\, ds(y), \quad x \in \partial S, \tag{A23}$$

where $\varphi \in C^{0,\alpha}(\partial S)$, $\alpha \in (0,1]$. Also, consider the functions

$$f^+(x) = \begin{cases} f(x), & x \in S_0^+, \\ p(x) + f_0(x), & x \in \partial S, \end{cases}$$
$$f^-(x) = \begin{cases} f(x), & x \in S_0^-, \\ -p(x) + f_0(x), & x \in \partial S, \end{cases} \tag{A24}$$

and
$$F^+(x) = \begin{cases} F(x), & x \in S_0^+, \\ p(x)\varphi(x) + F_0(x), & x \in \partial S, \end{cases}$$
$$F^-(x) = \begin{cases} F(x), & x \in S_0^-, \\ -p(x)\varphi(x) + F_0(x), & x \in \partial S, \end{cases} \tag{A25}$$

where $p \in C^{0,\alpha}(\partial S)$. If $f^+ \in C^{0,\alpha}(\bar{S}_0^+)$ and $f^- \in C^{0,\alpha}(\bar{S}_0^-)$, then $F^+ \in C^{0,\beta}(\bar{S}_0^+)$ and $F^- \in C^{0,\beta}(\bar{S}_0^-)$, with $\beta = \alpha$ for $\alpha \in (0,1)$ and any $\beta \in (0,1)$ for $\alpha = 1$.

Proof. From the properties of $k(x,y)$ it is clear that f_0 and F_0 are improper integrals. To prove the statement for F^+, it suffices to consider $x, x' \in \bar{S}_0^+$ satisfying (A12). Let $x = \xi + \sigma\nu(\xi) \in S_0^+$, $\xi \in \partial S$, and $x' = \xi' \in \partial S$. Then

$$\int_{\partial S} k(x,y)\varphi(y)\,ds(y) - p(x')\varphi(x') - \int_{\partial S} k(x',y)\varphi(y)\,ds(y)$$
$$= \int_{\partial S} k(x,y)\big[\varphi(y) - \varphi(\xi)\big]\,ds(y) - \int_{\partial S} k(x',y)\big[\varphi(y) - \varphi(x')\big]\,ds(y)$$
$$+ \big[\varphi(\xi) - \varphi(x')\big] \int_{\partial S} k(x,y)\,ds(y)$$
$$+ \varphi(x')\bigg(\int_{\partial S} k(x,y)\,ds(y) - p(x') - \int_{\partial S} k(x',y)\,ds(y)\bigg); \tag{A26}$$

that is,

$$F^+(x) - F^+(x') = \Phi(x) - \Phi(x') + \big[\varphi(\xi) - \varphi(\xi')\big]f^+(x) + \big[f^+(x) - f^+(x')\big]\varphi(\xi'), \tag{A27}$$

where Φ is given by (A17). The equality (A27) is similarly obtained when $x, x' \in S_0^+$, $x, x' \in \partial S$, or $x \in \partial S$, $x' \in S_0^+$. Since, by our assumption, both f_0 and f are bounded, (A22) shows that $k(x,y)$ satisfies estimate (A18). The assertion now follows from (A27) and Theorem A30.

F^- is treated analogously.

This theorem can be generalized to certain 1-singular kernels on ∂S.

A39. Definition. A 1-singular kernel on ∂S is called *integrable* if $\int_{\partial S} k(x,y)\,ds(y)$ exists as principal value for all $x \in \partial S$, and *uniformly integrable* if the integral in (A20) converges uniformly with respect to $x \in \partial S$.

For convenience, we extend this concept to γ-singular kernels, $\gamma \in (0,1)$, and note that all such kernels are uniformly integrable.

A40. Remark. If the kernel $k(x,y)$ is uniformly integrable, then $\int_{\partial S} k(x,y)\,ds(y)$ is continuous on ∂S. This is shown by writing the principal value of the integral as the sum of a uniformly convergent infinite series. Evidently, any uniformly integrable kernel satisfies (A18) on ∂S.

A41. Theorem. *If $k(x,y)$ is 1-singular on ∂S and integrable, and if $\varphi \in C^{0,\alpha}(\partial S)$, $\alpha \in (0,1]$, then the integral*

$$\int_{\partial S} k(x,y)\varphi(y)\,ds(y)$$

exists in the sense of principal value for all $x \in \partial S$. If $k(x,y)$ is uniformly integrable, then the above principal value exists uniformly with respect to $x \in \partial S$.

Proof. We write

$$\int_{\partial S \setminus \Sigma_{x,\delta}} k(x,y)\varphi(y)\,ds(y) = \int_{\partial S \setminus \Sigma_{x,\delta}} k(x,y)\bigl[\varphi(y) - \varphi(x)\bigr]\,ds(y)$$

$$+ \varphi(x) \int_{\partial S \setminus \Sigma_{x,\delta}} k(x,y)\,ds(y).$$

The result follows from the fact that, as $\delta \to 0$, the first term on the right-hand side converges uniformly since its integrand is $O(|x-y|^{\alpha-1})$.

A42. Theorem. *Suppose that*

(i) *$k(x,y)$ is a proper 1-singular kernel in S_0 that is integrable on ∂S;*

(ii) *f^+ and f^- defined by (A24), where $p \in C^{0,\alpha}(\partial S)$, $\alpha \in (0,1]$, and f_0 is understood as principal value, belong, respectively, to $C^{0,\alpha}(\bar{S}_0^+)$ and $C^{0,\alpha}(\bar{S}_0^-)$.*

Then the functions F^+ and F^- defined by (A25), where $\varphi \in C^{0,\alpha}(\partial S)$ and F_0 is understood as principal value, belong, respectively, to $C^{0,\beta}(\bar{S}_0^+)$ and $C^{0,\beta}(\bar{S}_0^-)$ with $\beta = \alpha$ for $\alpha \in (0,1)$ and any $\beta \in (0,1)$ for $\alpha = 1$.

Proof. By Theorem A41, F_0 exists in the sense of principal value for all $x \in \partial S$.

As in the proof of Theorem A38, let $x, x' \in \bar{S}_0^+$, $x \neq x'$. If $x, x' \in S_0^+$, then equality (A27) is established immediately. If $x \in S_0^+$, $x' \in \partial S$ (or $x \in \partial S$, $x' \in S_0^+$), we write (A26) with the integrals extended over $\partial S \setminus \Sigma_{x',\delta}$ ($\partial S \setminus \Sigma_{x,\delta}$) in the first instance, then let $\delta \to 0$. Noting that the limit of the second term on the right-hand side coincides with the improper integral $\Phi(x')$ $(\Phi(x))$, we again arrive at (A27). Finally, we see that this is also true if both $x, x' \in \partial S$, when the integrals in (A26) are initially extended over $\partial S \setminus (\Sigma_{x,\delta} \cup \Sigma_{x',\delta})$. Hence, (A27) holds for all $x, x' \in \bar{S}_0^+$, $x \neq x'$, and the result follows from assumptions (i) and (ii) and Theorem A30.

The reasoning is similar in the case of $\bar{S}_0^-$.

A4. Potential-type functions

In what follows we examine the Hölder continuity and continuous differentiability on $\bar{S}^+$ and $\bar{S}^-$ of functions that are analytic in S^+ and S^-. Hence, it suffices to consider the behaviour of such functions in the boundary layer S_0.

We begin by giving a brief account of the main properties of the functions

$$(v\varphi)(x) = -\int_{\partial S} (\ln|x-y|)\varphi(y)\,ds(y), \tag{A28}$$

$$(w\varphi)(x) = -\int_{\partial S} (\partial_{\nu(y)} \ln|x-y|)\varphi(y)\,ds(y) \tag{A29}$$

mentioned in Theorem 1.4.

A43. Theorem. *If $\varphi \in C(\partial S)$, then $v\varphi \in C^{0,\alpha}(\mathbb{R}^2)$ for any $\alpha \in (0,1)$.*

Proof. The assertion follows from Theorem A28 in view of the fact that, as can easily be verified by means of Lemma A23, the kernel $k(x,y) = -\ln|x-y|$ of v is a proper γ-singular kernel in S_0 for any $\gamma \in (0,1)$.

A44. Theorem. *If $\varphi \in C^{0,\alpha}(\partial S)$, $\alpha \in (0,1]$, then the restrictions of $w\varphi$ to S^+ and S^- have $C^{0,\beta}$-extensions to $\bar{S}^+$ and $\bar{S}^-$, respectively, with $\beta = \alpha$ for $\alpha \in (0,1)$ and any $\beta \in (0,1)$ for $\alpha = 1$. These extensions are given by*

$$(w\varphi)^+(x) = \begin{cases} (w\varphi)^+(x), & x \in S^+, \\ -\pi\varphi(x) + (w_0\varphi)(x), & x \in \partial S, \end{cases}$$
$$(w\varphi)^-(x) = \begin{cases} w(x), & x \in S^-, \\ \pi\varphi(x) + (w_0\varphi)(x), & x \in \partial S, \end{cases} \tag{A30}$$

where

$$(w_0\varphi)(x) = -\int_{\partial S} (\partial_{\nu(y)} \ln|x-y|)\varphi(y)\,ds(y), \quad x \in \partial S. \tag{A31}$$

Proof. Applying Lemmas A23 and A1, we can easily verify that

$$k(x, y) = -\partial_{\nu(y)} \ln |x - y| = \langle \nu(y), x - y \rangle |x - y|^{-2}$$

is a proper 1-singular kernel in S_0 and 0-singular on ∂S. Consequently, $w_0 \varphi$ is an improper integral.

Let $x \in S^+$, and consider a disk $\sigma_{x,\delta} \subset S^+$ with the centre at x and radius δ sufficiently small so that $\bar{\sigma}_{x,\delta}$ lies entirely in S^+. Using the divergence theorem in $\bar{S}^+ \setminus \sigma_{x,\delta}$ and the fact that $\ln |x - y|$ is a solution of the Laplace equation for $x \neq y$, we find that

$$\begin{aligned} 0 &= \int_{S^+ \setminus \sigma_{x,\delta}} \Delta(y) \ln |x - y| \, da(y) \\ &= \left(\int_{\partial S} - \int_{\partial \sigma_{x,\delta}} \right) \partial_{\nu(y)} \ln |x - y| \, ds(y) \\ &= \int_{\partial S} \partial_{\nu(y)} \ln |x - y| \, ds(y) - 2\pi, \end{aligned}$$

where $\partial \sigma_{x,\delta}$ is the circular boundary of $\sigma_{x,\delta}$; hence,

$$\int_{\partial S} \partial_{\nu(y)} \ln |x - y| \, ds(y) = 2\pi, \quad x \in S^+. \tag{A32}$$

The procedure is similar for $x \in \partial S$, except that in this case $\sigma_{x,\delta}$ is replaced by $\sigma_{x,\delta} \cap S^+$ and $\partial \sigma_{x,\delta}$ by its part lying in S^+. It is not difficult to show that, for δ small, the length of this part is equal to $\pi \delta + O(\delta^2)$, which leads to

$$\int_{\partial S} \partial_{\nu(y)} \ln |x - y| \, ds(y) = \pi, \quad x \in \partial S. \tag{A33}$$

Finally, the direct application of the divergence theorem yields

$$\int_{\partial S} \partial_{\nu(y)} \ln |x - y| \, ds(y) = 0, \quad x \in S^-. \tag{A34}$$

In view of these integrals and the expression of $k(x, y)$ we now see that

$$f(x) = \int_{\partial S} k(x, y) \, ds(y) = \begin{cases} -2\pi, & x \in S^+, \\ 0, & x \in S^-, \end{cases}$$

$$f_0(x) = \int_{\partial S} k(x, y) \, ds(y) = -\pi, \quad x \in \partial S. \tag{A35}$$

From (A24) with $p(x) = -\pi$, $x \in \partial S$, we obtain

$$f^+(x) = -2\pi, \quad x \in \bar{S}_0^+,$$
$$f^-(x) = 0, \quad x \in \bar{S}_0^-.$$

Since $f^+ \in C^{0,\alpha}(\bar{S}_0^+)$ and $f^- \in C^{0,\alpha}(\bar{S}_0^-)$, the result follows from Theorem A38.

A45. Remark. Theorem A44 implies that if φ belongs to $C^{0,\alpha}(\partial S)$, $\alpha \in (0,1]$, then, as $S^{\pm} \ni x' \to x \in \partial S$, $w\varphi$ tends to finite limits given by

$$(w\varphi)^{\pm}(x) = \mp \pi \varphi(x) - \int_{\partial S} (\partial_{\nu(y)} \ln|x-y|) \varphi(y)\, ds(y), \quad x \in \partial S, \qquad (A36)$$

where the last term is an improper integral. It can be shown [2] that $w\varphi$ can also be extended by continuity to $\bar{S}^+$ and $\bar{S}^-$ if $\varphi \in C(\partial S)$, but then the two extensions $(w\varphi)^+$ and $(w\varphi)^-$ are merely continuous.

A46. Theorem. *If $\varphi \in C^{0,\alpha}(\partial S)$, $\alpha \in (0,1]$, then the first order derivatives of $v\varphi$ in S^+ and S^- have $C^{0,\beta}$-extensions to $\bar{S}^+$ and $\bar{S}^-$, respectively, with $\beta = \alpha$ for $\alpha \in (0,1)$ and any $\beta \in (0,1)$ for $\alpha = 1$. These extensions are given by*

$$\bigl(\mathrm{grad}(v\varphi)\bigr)^+(x) = \begin{cases} \bigl(\mathrm{grad}(v\varphi)\bigr)(x), & x \in S^+, \\ \pi \nu(x)\varphi(x) + \bigl(\mathrm{grad}(v\varphi)\bigr)_0(x), & x \in \partial S, \end{cases}$$

$$\bigl(\mathrm{grad}(v\varphi)\bigr)^-(x) = \begin{cases} \bigl(\mathrm{grad}(v\varphi)\bigr)(x), & x \in S^-, \\ -\pi \nu(x)\varphi(x) + \bigl(\mathrm{grad}(v\varphi)\bigr)_0(x), & x \in \partial S, \end{cases}$$

where

$$\bigl(\mathrm{grad}(v\varphi)\bigr)_0(x) = -\int_{\partial S} \bigl(\mathrm{grad}(x) \ln|x-y|\bigr) \varphi(y)\, ds(y), \quad x \in \partial S,$$

the integral being understood as principal value.

Proof. By checking the properties required in Lemma A23, we convince ourselves that $k(x,y) = -\mathrm{grad}(x) \ln|x-y|$ is a proper 1-singular kernel in S_0 and on ∂S.

From (A15) and the fact that

$$\bigl(\mathrm{grad}(x) + \mathrm{grad}(y)\bigr) \ln|x-y| = 0, \quad x \neq y,$$

it follows that

$$k(x,y) = (\partial_{s(y)} \ln|x-y|)\tau(y) + (\partial_{\nu(y)} \ln|x-y|)\nu(y), \quad x \neq y.$$

Consequently, using integration by parts and denoting by a and b the end-points of the arc $\Sigma_{x,\delta}$, for $x \in \partial S$ we can write

$$\int_{\partial S \setminus \Sigma_{x,\delta}} k(x,y)\, ds(y)$$
$$= \int_{\partial S \setminus \Sigma_{x,\delta}} (\partial_{s(y)} \ln|x-y|)\tau(y)\, ds(y) + \int_{\partial S \setminus \Sigma_{x,\delta}} (\partial_{\nu(y)} \ln|x-y|)\nu(y)\, ds(y)$$
$$= [\tau(a) - \tau(b)]\ln \delta - \int_{\partial S \setminus \Sigma_{x,\delta}} (\ln|x-y|)\kappa(y)\nu(y)\, ds(y)$$
$$+ \int_{\partial S \setminus \Sigma_{x,\delta}} (\partial_{\nu(y)} \ln|x-y|)\nu(y)\, ds(y).$$

Since ∂S is a C^2-curve, the first term on the right-hand side vanishes as $\delta \to 0$, while the other two tend to $\bigl(v(\kappa\nu)\bigr)(x)$ and $-(w_0\nu)(x)$, respectively; therefore, $k(x,y)$ is integrable on ∂S and

$$f_0(x) = \int_{\partial S} k(x,y)\, ds(y) = \bigl(v(\kappa\nu)\bigr)(x) - (w_0\nu)(x), \quad x \in \partial S,$$

where f_0 is understood as principal value.

On the other hand, if $x \in S_0 \setminus \partial S$, then, again integrating by parts and taking (A1) into account, we find that

$$f(x) = \int_{\partial S} k(x,y)\, ds(y) = \bigl(v(\kappa\nu)\bigr)(x) - (w\nu)(x), \quad x \in S_0 \setminus \partial S.$$

By Theorems A43 and A44, the function f is $C^{0,\alpha}$-extendable to $\bar{S}_0^+$ and $\bar{S}_0^-$ and the values of the corresponding extensions on ∂S are given by the formula

$$f^\pm(x) = \bigl(v(\kappa\nu)\bigr)(x) \pm \pi\nu(x) + (w_0\nu)(x) = \pm\pi\nu(x) + f_0(x), \quad x \in \partial S;$$

in other words, these expressions are (A24) with $p = \pi\nu \in C^{0,1}(\partial S)$. As stated, $f^+ \in C^{0,\alpha}(\bar{S}_0^+)$ and $f^- \in C^{0,\alpha}(\bar{S}_0^-)$. The assertion now follows from Theorem A42 with F and F_0 in (A23) defined by

$$F(x) = -\int_{\partial S} (\mathrm{grad}(x) \ln|x-y|)\varphi(y)\, ds(y) = \bigl(\mathrm{grad}(v\varphi)\bigr)(x), \quad x \in S_0 \setminus \partial S,$$
$$F_0(x) = -\int_{\partial S} (\mathrm{grad}(x) \ln|x-y|)\varphi(y)\, ds(y) = \bigl(\mathrm{grad}(v\varphi)\bigr)_0(x), \quad x \in \partial S,$$

the latter understood as principal value.

A47. Remark. Theorem A46 implies that if $\varphi \in C^{0,\alpha}(\partial S)$, $\alpha \in (0,1]$, then, as $S^{\pm} \ni x' \to x \in \partial S$, $\mathrm{grad}(v\varphi)$ tends to finite limits given by

$$\big(\mathrm{grad}(v\varphi)\big)^{\pm}(x) = \pm\pi\nu(x)\varphi(x) - \int_{\partial S} \big(\mathrm{grad}(x)\ln|x-y|\big)\varphi(y)\,ds(y), \quad x \in \partial S, \quad \text{(A37)}$$

where the last term is understood as principal value.

A48. Remark. Theorems A46 and A17 also imply that if $\varphi \in C^{0,\alpha}(\partial S)$, $\alpha \in (0,1]$, then the restrictions of $v\varphi$ to $\bar{S}^+$ and $\bar{S}^-$ belong, respectively, to $C^{1,\beta}(\bar{S}^+)$ and $C^{1,\beta}(\bar{S}^-)$, with $\beta = \alpha$ for $\alpha \in (0,1)$ and any $\beta \in (0,1)$ for $\alpha = 1$. We denote these restrictions by $(v\varphi)^+$ and $(v\varphi)^-$; hence,

$$\begin{aligned}\big(\mathrm{grad}(v\varphi)^+\big)(x) &= \big(\mathrm{grad}(v\varphi)\big)^+(x), \quad x \in \bar{S}^+, \\ \big(\mathrm{grad}(v\varphi)^-\big)(x) &= \big(\mathrm{grad}(v\varphi)\big)^-(x), \quad x \in \bar{S}^-.\end{aligned}$$

A49. Theorem. *If $\varphi \in C^{1,\alpha}(\partial S)$, $\alpha \in (0,1]$, then the restrictions of $w\varphi$ to S^+ and S^- have $C^{1,\beta}$-extensions $(w\varphi)^+$ and $(w\varphi)^-$ to $\bar{S}^+$ and $\bar{S}^-$, respectively, with $\beta = \alpha$ for $\alpha \in (0,1)$ and any $\beta \in (0,1)$ for $\alpha = 1$. These extensions are given by (A30) and satisfy the equality $\partial_\nu(w\varphi)^+ = \partial_\nu(w\varphi)^-$ on ∂S.*

Proof. Let $x \neq y$. Since

$$\Delta(y)\ln|x-y| = 0, \quad \big(\mathrm{grad}(x) + \mathrm{grad}(y)\big)\ln|x-y| = 0,$$

we can write

$$\begin{aligned}\partial_{x_\gamma}\big(\partial_{\nu(y)}\ln|x-y|\big) &= \nu_\beta(y)\partial_{y_\beta}\big(\partial_{x_\gamma}\ln|x-y|\big) + \nu_\gamma(y)\Delta(y)\ln|x-y| \\ &= \nu_\beta(y)\partial_{y_\beta}(-\partial_{y_\gamma}\ln|x-y|) + \nu_\gamma(y)\partial_{y_\beta}(\partial_{y_\beta}\ln|x-y|) \\ &= \big(\nu_\beta(y)\partial_{y_\gamma} - \nu_\gamma(y)\partial_{y_\beta}\big)(\partial_{x_\beta}\ln|x-y|) = \varepsilon_{\beta\gamma}\partial_{s(y)}(\partial_{x_\beta}\ln|x-y|).\end{aligned}$$

Consequently, using integration by parts, we find that for $x \in S_0 \setminus \partial S$

$$\begin{aligned}\partial_{x_\gamma}(w\varphi)(x) &= -\int_{\partial S} \partial_{x_\gamma}\big(\partial_{\nu(y)}\ln|x-y|\big)\varphi(y)\,ds(y) \\ &= \varepsilon_{\beta\gamma}\partial_{x_\beta}\int_{\partial S}(\ln|x-y|)\varphi'(y)\,ds(y) = \varepsilon_{\gamma\beta}\partial_{x_\beta}(v\varphi')(x). \quad \text{(A38)}\end{aligned}$$

From this, Theorem A46 and the fact that $\varphi' \in C^{0,\alpha}(\partial S)$ we deduce that $\mathrm{grad}(w\varphi)$ has $C^{0,\beta}$-extensions $\big(\mathrm{grad}(w\varphi)\big)^+$ and $\big(\mathrm{grad}(w\varphi)\big)^-$ to $\bar{S}^+$ and $\bar{S}^-$, respectively. By

Theorem A44, the extensions $(w\varphi)^+$ and $(w\varphi)^-$, given by (A30), of $w\varphi$ are Hölder continuous on $\bar{S}^+$ and $\bar{S}^-$, respectively. Since $\bigl(\operatorname{grad}(w\varphi)^+\bigr)(x) = \bigl(\operatorname{grad}(w\varphi)\bigr)^+(x)$, $x \in S^+$, and $\bigl(\operatorname{grad}(w\varphi)^-\bigr)(x) = \bigl(\operatorname{grad}(w\varphi)\bigr)^-(x)$, $x \in S^-$, the first part of the assertion follows from Theorem A17.

To complete the proof, we remark that, in view of (A38) and (A37), for $x \in \partial S$ Theorem A17 yields

$$\begin{aligned}
\partial_\nu(w\varphi)^\pm(x) &= \bigl\langle \bigl(\operatorname{grad}(w\varphi)^\pm\bigr)(x), \nu(x) \bigr\rangle \\
&= \bigl\langle \bigl(\operatorname{grad}(w\varphi)^\pm\bigr)(x), \nu(x) \bigr\rangle \\
&= \varepsilon_{\gamma\beta}\bigl(\partial_{x_\beta}(v\varphi')\bigr)^\pm(x)\nu_\gamma(x) \\
&= \varepsilon_{\beta\gamma}\nu_\gamma(x)\int_{\partial S}(\partial_{x_\beta}\ln|x-y|)\varphi'(y)\,ds(y),
\end{aligned}$$

where the integral is understood as principal value.

A50. Theorem. *The function $w_0\varphi$ defined by (A31) as the direct values on ∂S of the double layer potential with density $\varphi \in C^{0,\alpha}(\partial S)$, $\alpha \in (0,1]$, belongs to $C^{1,\beta}(\partial S)$, with $\beta = \alpha$ for $\alpha \in (0,1)$ and any $\beta \in (0,1)$ for $\alpha = 1$.*

Proof. As noted in the proof of Theorem A44, the kernel $k(x,y) = -\partial_{\nu(y)}\ln|x-y|$ is 0-singular on ∂S; consequently, $(w_0\varphi)(x)$ is an improper integral for all $x \in \partial S$. On the other hand, the kernel

$$k_0(x,y) = \partial_{s(x)}k(x,y) = \frac{\langle \nu(y), \tau(x)\rangle}{|x-y|^2} - 2\frac{\langle \nu(y), x-y\rangle\langle \tau(x), x-y\rangle}{|x-y|^4} \quad (A39)$$

is 1-singular on ∂S. Verifying the conditions of Lemma A23, we deduce that $k_0(x,y)$ is a proper 1-singular kernel on ∂S.

Next, by writing $\langle\cdot,\cdot\rangle$ in terms of the cosine of the angle between the vectors, we find that

$$\langle \nu(y), \tau(x)\rangle + \langle \nu(x), \tau(y)\rangle = 0, \quad x, y \in \partial S. \quad (A40)$$

Using the same technique, (A39) and (A40), for $x, y \in \partial S$, $x \neq y$, we now obtain

$$\begin{aligned}
(\partial_{s(x)} + \partial_{s(y)})k(x,y) = 2|x-y|^{-4}\bigl[&\langle \nu(x), x-y\rangle\langle \tau(y), x-y\rangle \\
&- \langle \nu(y), x-y\rangle\langle \tau(x), x-y\rangle + \langle \nu(y), \tau(x)\rangle\bigr] = 0. \quad (A41)
\end{aligned}$$

From this and (A39) we conclude that $k_0(x,y)$ satisfies (A18).

The assertion now follows from Theorem A32 with $\beta = \gamma = 0$ and $g(x) = -\pi$, $x \in \partial S$ (according to (A35)), and Remarks A34 and A33.

A5. Other potential-type functions

We turn our attention to the Hölder continuity and continuous differentiability of the other functions with γ-singular kernels mentioned in Theorems 1.3 and 1.4. For convenience, the various parts of these theorems are given as separate assertions. Also, as already observed, it suffices to establish the necessary properties in S_0.

A51. Theorem. *Let $k(x,y)$ be continuous in $S_0 \times \partial S$ and such that $\mathrm{grad}(x)k(x,y)$ is a proper γ-singular kernel in S_0, $\gamma \in [0,1)$, and let*

$$(v^a\varphi)(x) = \int_{\partial S} k(x,y)\varphi(y)\,ds(y), \quad x \in S_0.$$

If $\varphi \in C(\partial S)$, then $v^a\varphi \in C^{1,\beta}(S_0)$, with $\beta = 1 - \gamma$ for $\gamma \in (0,1)$ and any $\beta \in (0,1)$ for $\gamma = 0$.

Proof. Clearly, $v^a\varphi \in C(S_0) \cap C^1(S_0^+) \cap C^1(S_0^-)$. The statement follows from the fact that for $x \in S_0 \setminus \partial S$,

$$\big(\mathrm{grad}(v^a\varphi)\big)(x) = \int_{\partial S} \mathrm{grad}(x)k(x,y)\varphi(y)\,ds(y),$$

which, by Theorem A28, belongs to $C^{0,\alpha}(S_0)$.

A52. Theorem. *Let $k(x,y)$ be continuous on $\partial S \times \partial S$ and such that $\partial_{s(x)}k(x,y)$ is a proper γ-singular kernel on ∂S, $\gamma \in [0,1)$. If $\varphi \in C(\partial S)$, then the function*

$$(v_0^a\varphi)(x) = \int_{\partial S} k(x,y)\varphi(y)\,ds(y), \quad x \in \partial S,$$

belongs to $C^{1,\beta}(\partial S)$, with $\beta = 1 - \gamma$ for $\gamma \in (0,1)$ and any $\beta \in (0,1)$ for $\gamma = 0$.

Proof. Consider the function

$$(v_{0\delta}^a\varphi)(x) = \int_{\partial S \setminus \Sigma_{x,\delta}} k(x,y)\varphi(y)\,ds(y), \quad \delta > 0.$$

It is obvious that $(v_{0\delta}^a\varphi)(x) \to (v_0^a\varphi)(x)$ as $\delta \to 0$, for all $x \in \partial S$. On the other hand,

$$\partial_s(v_{0\delta}^a\varphi)(x) = \int_{\partial S \setminus \Sigma_{x,\delta}} \partial_{s(x)}k(x,y)\varphi(y)\,ds(y),$$

which converges uniformly to $\int_{\partial S} \partial_{s(x)} k(x,y)\varphi(y)\,ds(y)$ as $\delta \to 0$ (see the proof of Theorem A27). By a well-known theorem of analysis, $v_0^a \varphi$ is differentiable at all $x \in \partial S$ and

$$\partial_s(v_0^a \varphi)(x) = \int_{\partial S} \partial_{s(x)} k(x,y)\varphi(y)\,ds(y).$$

We complete the proof by applying Theorem A28 to the above integral to deduce that $\partial_s(v_0^a \varphi) \in C^{0,\beta}(\partial S)$.

A53. Theorem. *If $\varphi \in C(\partial S)$, then the functions*

$$(v_{\gamma\delta}^b \varphi)(x) = \int_{\partial S} \frac{(x_\gamma - y_\gamma)(x_\delta - y_\delta)}{|x-y|^2} \varphi(y)\,ds(y), \quad x \in S_0, \tag{A42}$$

$$(v_\gamma^c \varphi)(x) = \int_{\partial S} \left[\partial_{s(y)}\big((x_\gamma - y_\gamma)\ln|x-y|\big)\right]\varphi(y)\,ds(y), \quad x \in S_0, \tag{A43}$$

$$(v_\gamma^d \varphi)(x) = \int_{\partial S} \left[\partial_{\nu(y)}\big((x_\gamma - y_\gamma)\ln|x-y|\big)\right]\varphi(y)\,ds(y), \quad x \in S_0, \tag{A44}$$

belong to $C^{0,\alpha}(S_0)$ for any $\alpha \in (0,1)$.

Proof. By direct verification or by means of Lemma A23, we easily convince ourselves that $(x_\gamma - y_\gamma)(x_\delta - y_\delta)|x-y|^{-2}$ is a proper 0-singular kernel in S_0. Similarly,

$$\partial_{s(y)}\big((x_\gamma - y_\gamma)\ln|x-y|\big) = -\tau_\gamma \ln|x-y| - \frac{(x_\gamma - y_\gamma)\langle \tau(y), x-y\rangle}{|x-y|^2}$$

and

$$\partial_{\nu(y)}\big((x_\gamma - y_\gamma)\ln|x-y|\big) = -\nu_\gamma(y)\ln|x-y| - \frac{(x_\gamma - y_\gamma)\langle \nu(y), x-y\rangle}{|x-y|^2}$$

are proper σ-singular kernels in S_0 for any $\sigma \in (0,1)$. The result now follows from Theorem A28.

A54. Theorem. *If $\varphi \in C^{0,\alpha}(\partial S)$, $\alpha \in (0,1]$, then the function*

$$(v_{\gamma\delta}^e \varphi)(x) = \int_{\partial S} \left(\partial_{s(y)} \frac{(x_\gamma - y_\gamma)(x_\delta - y_\delta)}{|x-y|^2}\right)\varphi(y)\,ds(y), \quad x \in S_0, \tag{A45}$$

belongs to $C^{0,\beta}(S_0)$, with $\beta = \alpha$ for $\alpha \in (0,1)$ and any $\beta \in (0,1)$ for $\alpha = 1$.

Proof. The direct verification of the properties in Definition A16 shows that the kernel $k(x,y)$ of the operator $v_{\gamma\delta}^e$ is a proper 1-singular kernel in S_0. Also, for any $x, y \in \partial S$, $x \neq y$,

$$\partial_{s(y)} \frac{(x_\gamma - y_\gamma)(x_\delta - y_\delta)}{|x-y|^2} = c_{\gamma\delta\rho\sigma} \frac{(x_\rho - y_\rho)(x_\sigma - y_\sigma)}{|x-y|^2} \partial_{\nu(y)} \ln|x-y|, \tag{A46}$$

where

$$\begin{aligned} c_{\gamma\gamma\gamma\delta} &= c_{\gamma\gamma\delta\gamma} = -c_{\gamma\delta\gamma\gamma} = -c_{\delta\gamma\gamma\gamma} = \varepsilon_{\delta\gamma}, \\ c_{\gamma\delta\gamma\delta} &= c_{\gamma\delta\delta\gamma} = c_{\gamma\gamma\delta\delta} = 0 \quad (\gamma, \delta \text{ not summed}), \end{aligned} \tag{A47}$$

which means that $k(x,y)$ is 0-singular on ∂S. Consequently, $v_{\gamma\delta}^e \varphi$ is an improper integral for $x \in \partial S$.

Since for $x, y \in \partial S$, $x \neq y$,

$$\lim_{y \to x} \frac{(x_\gamma - y_\gamma)(x_\delta - y_\delta)}{|x-y|^2} = \tau_\gamma(x) \tau_\delta(x),$$

we find that f and f_0 defined by (A22) are identically zero. Hence, f^+ and f^- defined by (A24) with $p(x) = 0$, $x \in \partial S$, belong to $C^{0,\alpha}(\partial S)$. The result now follows from Theorem A38.

A55. Theorem. *If $\varphi \in C^{0,\alpha}(\partial S)$, $\alpha \in (0,1]$, then*

$$(v_0^f \varphi)(x) = \int_{\partial S} \left(\partial_{s(y)} \ln|x-y| \right) \varphi(y) \, ds(y), \quad x \in \partial S, \tag{A48}$$

exists as principal value uniformly for all $x \in \partial S$. Furthermore, $v_0^f \varphi \in C^{0,\beta}(\partial S)$, with $\beta = \alpha$ for $\alpha \in (0,1)$ and any $\beta \in (0,1)$ for $\alpha = 1$.

Proof. For $x, y \in \partial S$, $x \neq y$, we have

$$|\partial_{s(y)} \ln|x-y|| = \frac{|\langle \tau(y), x-y \rangle|}{|x-y|^2} \leq c_1 |x-y|^{-1},$$

$$|x-y| \, |\partial_{x_\gamma} (\partial_{s(y)} \ln|x-y|)| = \left| \frac{\tau_\gamma(y)}{|x-y|} - 2 \frac{\langle \tau(y), x-y \rangle (x_\gamma - y_\gamma)}{|x-y|^3} \right| \leq c_2 |x-y|^{-1},$$

where c_1 and c_2 are positive constants; therefore, by Lemma A23, $\partial_{s(y)} \ln|x-y|$ is a proper 1-singular kernel on ∂S. This kernel is also uniformly integrable since if a and b are the end-points of $\Sigma_{x,\delta}$, then

$$\int_{\partial S \setminus \Sigma_{x,\delta}} \partial_{s(y)} \ln|x-y| \, ds(y) = \ln \frac{|x-a|}{|x-b|} = 0 \tag{A49}$$

for all $0 < \delta \leq r$ and all $x \in \partial S$. We can now write

$$\int_{\partial S \setminus \Sigma_{x,\delta}} \left(\partial_{s(y)} \ln|x-y|\right) \varphi(y) \, ds(y) = \int_{\partial S \setminus \Sigma_{x,\delta}} \left(\partial_{s(y)} \ln|x-y|\right) \left[\varphi(y) - \varphi(x)\right] ds(y),$$

and the first part of the assertion follows from Definition A35 and the uniform convergence, as $\delta \to 0$, of the right-hand side, whose integrand is $O(|x-y|^{\alpha-1})$. Consequently,

$$(v_0^f \varphi)(x) = \int_{\partial S} \left(\partial_{s(y)} \ln|x-y|\right) \left[\varphi(y) - \varphi(x)\right] ds(y), \quad x \in \partial S, \qquad \text{(A50)}$$

in the sense of principal value.

To complete the proof, we apply Theorem A30 with $\xi \equiv x$ and make use of the last part of Remark A40.

A56. Theorem. *If $\varphi \in C^{0,\alpha}(\partial S)$, $\alpha \in (0,1]$, then the function*

$$(v^f \varphi)(x) = \int_{\partial S} \left(\partial_{s(y)} \ln|x-y|\right) \varphi(y) \, ds(y), \quad x \in S_0 \setminus \partial S, \qquad \text{(A51)}$$

is $C^{0,\beta}$-extendable to $\mathbb{R}^2$, with $\beta = \alpha$ for $\alpha \in (0,1)$ and any $\beta \in (0,1)$ for $\alpha = 1$.

Proof. In the proof of Theorem A55 it was shown that $k(x,y) = \partial_{s(y)} \ln|x-y|$ is an integrable, proper 1-singular kernel on ∂S. The same reasoning indicates that $k(x,y)$ is also a proper 1-singular kernel in S_0. In view of (A49), formulae (A22) yield

$$\begin{aligned} f(x) &= 0, \quad x \in S_0 \setminus \partial S, \\ f_0(x) &= 0, \quad x \in \partial S, \end{aligned} \qquad \text{(A52)}$$

the latter understood as principal value. From (A52) and (A24) with $p(x) = 0$, $x \in \partial S$, it follows that $f^+ \in C^{0,\alpha}(\bar{S}_0^+)$ and $f^- \in C^{0,\alpha}(\bar{S}_0^-)$ (both these functions are identically zero). The application of Theorem A42 now completes the proof.

A57. Remark. Since $p = 0$, (A51) also represents the extension of $v^f \varphi$ to $\mathbb{R}^2$, that is, it holds for $x \in \mathbb{R}^2$, but for $x \in \partial S$ the integral on the right-hand side (denoted by $v_0^f \varphi$ in (A48)) must be understood as principal value.

Alternatively, since

$$\int_{\partial S} \partial_{s(y)} \ln|x-y| \, ds(y) = 0, \quad x \in \mathbb{R}^2 \setminus \partial S,$$

we see that the extension of $v^f\varphi$ to $\mathbb{R}^2$ is also given by the right-hand side of (A50) with $x \in \mathbb{R}^2$.

A58. Theorem. *If $\varphi \in C^{1,\alpha}(\partial S)$, $\alpha \in (0,1]$, then the function $v_0^f\varphi$ defined by (A48) belongs to $C^{1,\beta}(\partial S)$, with $\beta = \alpha$ for $\alpha \in (0,1)$ and any $\beta \in (0,1)$ for $\alpha = 1$.*

Proof. By Theorem A55, $v_0^f\varphi$ is Hölder continuous on ∂S.

Let $x = \psi(s) \in \partial S$ be arbitrary but fixed, and let $a = \psi(s-\delta)$ and $b = \psi(s+\delta)$ be the end-points of the arc $\Gamma_{x,\delta}$ defined by (A21). Integrating by parts, we find that

$$\int_{\partial S \setminus \Gamma_{x,\delta}} (\ln|x-y|)\varphi'(y)\,ds(y) = \varphi(a)\ln|x-a| - \varphi(b)\ln|x-b|$$
$$- \int_{\partial S \setminus \Gamma_{x,\delta}} (\partial_{s(y)} \ln|x-y|)\varphi(y)\,ds(y). \qquad (A53)$$

The first term on the right-hand side can be written in the form

$$\varphi(x)(\ln|x-a| - \ln|x-b|) + [\varphi(a) - \varphi(x)]\ln|x-a| - [\varphi(b) - \varphi(x)]\ln|x-b|.$$

Since

$$\ln|x-a| - \ln|x-b| = \ln\left(\frac{|x-a|}{\delta} \cdot \frac{\delta}{|x-b|}\right)$$

and φ is differentiable on ∂S, it follows that, by Theorem A14, this expression tends to zero as $\delta \to 0$.

In the proof of Theorem A55 it was shown that $\partial_{s(y)} \ln|x-y|$ is an integrable, proper 1-singular kernel on ∂S. Setting

$$F(x) = \int_{\partial S} (\ln|x-y|)\varphi'(y)\,ds(y), \quad F_\delta(x) = \int_{\partial S \setminus \Gamma_{x,\delta}} (\ln|x-y|)\varphi'(y)\,ds(y)$$

and passing to the limit as $\delta \to 0$ in (A53), we see that, by Theorem A55 and Remark A36,

$$F(x) = \lim_{\delta \to 0} F_\delta(x) = -(v_0^f\varphi)(x). \qquad (A54)$$

On the other hand, by Leibniz's rule for differentiating an integral whose limits depend on the differentiation variable,

$$F'_\delta(x) = \int_{\partial S \setminus \Gamma_{x,\delta}} (\partial_{s(x)} \ln|x-y|)\varphi'(y)\,ds(y) + \varphi'(a)\ln|x-a| - \varphi'(b)\ln|x-b|.$$

Since $\varphi' \in C^{0,\alpha}(\partial S)$, we deduce as above that the sum of the last two terms tends to zero uniformly as $\delta \to 0$. Hence,

$$\lim_{\delta \to 0} F'_\delta(x) = \int_{\partial S} (\partial_{s(x)} \ln |x-y|) \varphi'(y) \, ds(y), \qquad (A55)$$

where the integral is understood as principal value and, by Theorem A55, the convergence is uniform with respect to x. A well-known result of analysis now implies that $F(x)$ is differentiable and that $F'(x)$ is equal to the right-hand side in (A55). Taking (A54) into account, we conclude that $\partial_s(v_0^f \varphi)(x)$ exists and

$$\partial_s(v_0^f \varphi)(x) = -\int_{\partial S} (\partial_{s(x)} \ln |x-y|) \varphi'(y) \, ds(y), \quad x \in \partial S.$$

By Theorem 1.57, $\partial_s(v_0^f \varphi) \in C^{0,\alpha}(\partial S)$, as required.

A59. Theorem. *If $\varphi \in C^{0,\alpha}(\partial S)$, $\alpha \in (0,1]$, then the functions*

$$(v_{\gamma 0}^c \varphi)(x) = \int_{\partial S} \left[\partial_{s(y)} \big((x_\gamma - y_\gamma) \ln |x-y| \big) \right] \varphi(y) \, ds(y), \quad x \in \partial S, \qquad (A56)$$

$$(v_{\gamma 0}^d \varphi)(x) = \int_{\partial S} \left[\partial_{\nu(y)} \big((x_\gamma - y_\gamma) \ln |x-y| \big) \right] \varphi(y) \, ds(y), \quad x \in \partial S, \qquad (A57)$$

belong to $C^{1,\beta}(\partial S)$, with $\beta = \alpha$ for $\alpha \in (0,1)$ and any $\beta \in (0,1)$ for $\alpha = 1$.

Proof. The kernel

$$k(x,y) = \partial_{s(y)}\big((x_\gamma - y_\gamma) \ln |x-y|\big) = -\tau_\gamma(y) \ln |x-y| - \frac{(x_\gamma - y_\gamma)\langle \tau(y), x-y \rangle}{|x-y|^2}$$

is δ-singular on ∂S, where $\delta \in (0,1)$ is arbitrary, so $(v_{\gamma 0}^c \varphi)(x)$ is an improper integral for all $x \in \partial S$. Also, the kernel

$$k_0(x,y) = \partial_{s(x)} k(x,y)$$
$$= -\tau_\gamma(y) \partial_{s(x)} \ln |x-y| - \partial_{s(x)} \frac{(x_\gamma - y_\gamma)\langle \tau(y), x-y \rangle}{|x-y|^2} \qquad (A58)$$

is 1-singular on ∂S. Using Lemma A23, we find that $k_0(x,y)$ is a proper 1-singular kernel on ∂S.

Since

$$\hat{k}(x,y) = (\partial_{s(x)} + \partial_{s(y)}) \ln |x-y| = \frac{\langle \tau(x) - \tau(y), x-y \rangle}{|x-y|^2}$$

is 0-singular on ∂S, the first term on the right-hand side in (A58) can be written in the form

$$-\hat{k}(x,y)\tau_\gamma(y) + \partial_{s(y)}\bigl(\tau_\gamma(y)\ln|x-y|\bigr) + \kappa(y)\nu_\gamma(y)\ln|x-y|.$$

Similarly, since

$$\tilde{k}_{\gamma\delta}(x,y) = (\partial_{s(x)} + \partial_{s(y)})\frac{(x_\gamma - y_\gamma)(x_\delta - y_\delta)}{|x-y|^2}$$

$$= \frac{\bigl[\tau_\gamma(x) - \tau_\gamma(y)\bigr](x_\delta - y_\delta)}{|x-y|^2} + \frac{\bigl[\tau_\delta(x) - \tau_\delta(y)\bigr](x_\gamma - y_\gamma)}{|x-y|^2}$$

$$- 2\frac{(x_\gamma - y_\gamma)(x_\delta - y_\delta)\langle\tau(x) - \tau(y), x - y\rangle}{|x-y|^4}$$

is 0-singular on ∂S, the second term on the right-hand side in (A58) becomes

$$-\tilde{k}_{\gamma\delta}(x,y)\tau_\delta(y) + \partial_{s(y)}\frac{(x_\gamma - y_\gamma)\langle\tau(y), x - y\rangle}{|x-y|^2} + \frac{\kappa(y)(x_\gamma - y_\gamma)\langle\nu(y), x - y\rangle}{|x-y|^2}.$$

Denoting by a and b the end-points of $\Sigma_{x,\delta}$, we find that

$$\int_{\partial S\setminus\Sigma_{x,\delta}} \partial_{s(y)}\bigl(\tau_\gamma(y)\ln|x-y|\bigr)\,ds(y) = \tau_\gamma(a)\ln|x-a| - \tau_\gamma(b)\ln|x-b|$$

$$= \tau_\gamma(a)\ln\frac{|x-a|}{|x-b|} + \bigl[\tau_\gamma(a) - \tau_\gamma(b)\bigr]\ln|x-b|$$

$$= \bigl[\tau_\gamma(a) - \tau_\gamma(b)\bigr]\ln|x-b| \to 0$$

uniformly as $\delta \to 0$, and that

$$\int_{\partial S\setminus\Sigma_{x,\delta}} \partial_{s(y)}\frac{(x_\gamma - y_\gamma)\langle\tau(y), x - y\rangle}{|x-y|^2}\,ds(y)$$

$$= \frac{(x_\gamma - a_\gamma)\langle\tau(a), x - a\rangle}{|x-a|^2} - \frac{(x_\gamma - b_\gamma)\langle\tau(b), x - b\rangle}{|x-b|^2} \to 0$$

uniformly as $\delta \to 0$. Consequently, $k_0(x,y)$ satisfies estimate (A18).

The result now follows from Theorem A32 with any $\beta \in (0,\alpha)$, $\gamma = 0$, and $g(x) = 0$, $x \in \partial S$, and Remarks A34 and A33.

The function $v_{\gamma 0}^d\varphi$ is treated similarly.

OTHER POTENTIAL-TYPE FUNCTIONS

A60. Theorem. *If $\varphi \in C^{0,\alpha}(\partial S)$, $\alpha \in (0,1]$, then the function defined by*

$$(v^e_{\gamma\delta 0}\varphi)(x) = \int_{\partial S} \left(\partial_{s(y)} \frac{(x_\gamma - y_\gamma)(x_\delta - y_\delta)}{|x - y|^2} \right) \varphi(y)\, ds(y), \quad x \in \partial S, \quad (A59)$$

belongs to $C^{1,\beta}(\partial S)$, with $\beta = \alpha$ for $\alpha \in (0,1)$ and any $\beta \in (0,1)$ for $\alpha = 1$.

Proof. From the formula (A46) and the estimates in Lemma A1 we see that the kernel

$$k(x,y) = \partial_{s(y)} \frac{(x_\gamma - y_\gamma)(x_\delta - y_\delta)}{|x - y|^2}$$

is 0-singular on ∂S; hence, $(v^e_{\gamma\delta 0}\varphi)(x)$ is an improper integral for all $x \in \partial S$. Next, a simple calculation shows that

$$k_0(x,y) = \partial_{s(x)} k(x,y)$$

$$= c_{\gamma\delta\lambda\mu} c_{\lambda\mu\rho\sigma} \frac{(x_\rho - y_\rho)(x_\sigma - y_\sigma)}{|x-y|^2} (\partial_{\nu(x)} \ln|x-y|)(\partial_{\nu(y)} \ln|x-y|)$$

$$+ c_{\gamma\delta\rho\sigma} \frac{(x_\rho - y_\rho)(x_\sigma - y_\sigma)}{|x-y|^2} \partial_{s(x)}(\partial_{\nu(y)} \ln|x-y|), \quad (A60)$$

where the $c_{\gamma\delta\rho\sigma}$ are given by (A47), is a 1-singular kernel on ∂S. Moreover, using Lemma A23, we easily convince ourselves that $k_0(x,y)$ is a proper 1-singular kernel on ∂S.

The first term on the right-hand side in (A60) is 0-singular on ∂S. By (A41), the second term can be written in the form

$$c_{\gamma\delta\rho\sigma}\left[-\partial_{s(y)}\left(\frac{(x_\rho - y_\rho)(x_\sigma - y_\sigma)}{|x-y|^2} \partial_{\nu(y)} \ln|x-y| \right) \right.$$
$$\left. + (\partial_{\nu(y)} \ln|x-y|)\partial_{s(y)} \frac{(x_\rho - y_\rho)(x_\sigma - y_\sigma)}{|x-y|^2} \right],$$

from which, in view of what was said above about $k(x,y)$, we immediately deduce by direct verification that $k_0(x,y)$ satisfies estimate (A18).

The assertion now follows from Theorem A32 with $\beta = \gamma = 0$ and $g(x) = 0$, $x \in \partial S$, and Remarks A34 and A33.

A6. Complex singular kernels

Extending an earlier convention, for a function f given on ∂S we write $f(z) \equiv f(x)$, where $z = x_1 + ix_2$.

We now assume that $C(\partial S)$ and $C^1(\partial S)$ are complex vector spaces, and construct the complex spaces $C^{0,\alpha}(\partial S)$ and $C^{1,\alpha}(\partial S)$ by defining Hölder continuity in terms of the inequality

$$|f(z) - f(\zeta)| \leq c|z - \zeta|^\alpha \quad \text{for all } z, \zeta \in \partial S,$$

and the derivative as

$$f'(z) = \frac{d}{dz}f(z) = \lim_{\zeta \to z} \frac{f(\zeta) - f(z)}{\zeta - z}, \quad z, \zeta \in \partial S,$$

if this limit exists.

Since $|z - \zeta| = |x - y|$, where $\zeta = y_1 + iy_2$, it is obvious that Hölder continuity with respect to z and Hölder continuity with respect to x (or s, according to the discussion in §A2) are equivalent. The same can also be said about Hölder continuous differentiability on ∂S. We can see this from the equality $f'(s) = \vartheta(z)f'(z)$, where

$$\vartheta(z) = \frac{dz}{ds} = \tau_1(z) + i\tau_2(z), \quad \text{(A61)}$$

which implies that $f' \in C^{0,\alpha}(\partial S)$ in terms of z if and only if $f' \in C^{0,\alpha}(\partial S)$ in terms of s, in view of Lemma A19 and the fact that both $\vartheta(z)$ and

$$\bar{\vartheta}(z) = [\vartheta(z)]^{-1} = \tau_1(z) - i\tau_2(z)$$

belong to $C^1(\partial S)$. This shows that our somewhat loose use of the same symbol for a function on ∂S, whether it is expressed in terms of z or x, is justified in relation to Hölder spaces.

In the light of these arguments, and because for a kernel $k(x,y)$ and a density φ on ∂S

$$\int_{\partial S} k(x,y)\varphi(y)\,ds(y) = \int_{\partial S} k(z,\zeta)\varphi(\zeta)\bar{\vartheta}(\zeta)\,d\zeta,$$

we conclude that the definition of γ-singular and proper γ-singular kernels on ∂S, and all the associated results established in §A4 on the behaviour on ∂S of integrals with such kernels, can be understood in terms of either real or complex variables.

A61. Theorem. *If $\varphi \in C^{0,\alpha}(\partial S)$, $\alpha \in (0,1]$, then the integral*

$$\Psi(z) = \int_{\partial S} \frac{\varphi(\zeta)}{\zeta - z} d\zeta, \quad z \in \partial S, \tag{A62}$$

exists in the sense of principal value, uniformly for all $z \in \partial S$, and belongs to $C^{0,\beta}(\partial S)$, with $\beta = \alpha$ for $\alpha \in (0,1)$ and any $\beta \in (0,1)$ for $\alpha = 1$.

Proof. Let $z = x_1 + ix_2$ and $\zeta = y_1 + iy_2$. Differentiating with respect to $s(y)$ the equality

$$\log(\zeta - z) = \ln|\zeta - z| + i\theta = \ln|x - y| + i\theta,$$

where $\theta = \arg(\zeta - z)$, and using the Cauchy-Riemann relation

$$\partial_{s(y)}\theta(x,y) = \partial_{\nu(y)} \ln|x - y|,$$

we obtain

$$\frac{d\zeta}{\zeta - z} = \partial_{s(y)} \ln|x - y|\, ds(y) + i\partial_{\nu(y)} \ln|x - y|\, ds(y). \tag{A63}$$

Hence, we can write

$$\int_{\partial S \setminus \Sigma_{x,\delta}} \frac{\varphi(\zeta)}{\zeta - z} d\zeta = \int_{\partial S \setminus \Sigma_{x,\delta}} (\partial_{s(y)} \ln|x-y|)\varphi(y)\, ds(y) + i \int_{\partial S \setminus \Sigma_{x,\delta}} (\partial_{\nu(y)} \ln|x-y|)\varphi(y)\, ds(y),$$

and then establish the result from Theorems A55 and A44 by passing to the limit as $\delta \to 0$.

A62. Remark. The function Ψ defined by (A62) can be expressed in terms of an improper integral. Writing

$$\int_{\partial S \setminus \Sigma_{x,\delta}} \frac{\varphi(\zeta)}{\zeta - z} d\zeta = \int_{\partial S \setminus \Sigma_{x,\delta}} \frac{\varphi(\zeta) - \varphi(z)}{\zeta - z} d\zeta + \varphi(z) \int_{\partial S \setminus \Sigma_{x,\delta}} \frac{d\zeta}{\zeta - z},$$

replacing $(\zeta - z)^{-1} d\zeta$ by its expression in (A63), letting $\delta \to 0$ and using formulae (A49) and (A35), we find that, in the sense of principal value,

$$\int_{\partial S} \frac{\varphi(\zeta)}{\zeta - z} d\zeta = \pi i \varphi(z) + \int_{\partial S} \frac{\varphi(\zeta) - \varphi(z)}{\zeta - z} d\zeta, \quad z \in \partial S, \tag{A64}$$

where the integrand of the last term is $O(|z - \zeta|^{\alpha - 1})$ if $\varphi \in C^{0,\alpha}(\partial S)$, $\alpha \in (0,1]$.

A63. Theorem. *If $\varphi \in C^{1,\alpha}(\partial S)$, $\alpha \in (0,1]$, then Ψ defined by (A62) belongs to $C^{1,\beta}(\partial S)$, with $\beta = \alpha$ for $\alpha \in (0,1)$ and any $\beta \in (0,1)$ for $\alpha = 1$.*

Proof. By (A63), (A48) and (A31), $\Psi(z) = (v_0^f \varphi)(x) - i(w_0 \varphi)(x)$, and the assertion follows from Theorems A58 and A50.

A64. Theorem. *If $K^s : C^{0,\alpha}(\partial S) \to C^{0,\alpha}(\partial S)$, $\alpha \in (0,1)$, is the operator defined by*

$$(K^s \varphi)(z) = \int_{\partial S} \frac{\varphi(\zeta)}{\zeta - z} \, d\zeta, \quad z \in \partial S, \tag{A65}$$

then $(K^s)^2 = -\pi^2 I$, where I is the identity operator.

Proof. From Theorem A61 it is clear that the operator composition $(K^s)^2$ is meaningful.

In [20] it is shown that a function $f(z, \zeta)$ which is Hölder continuous with respect to both its variables z and ζ satisfies the Poincaré-Bertrand formula

$$\int_{\partial S} \frac{1}{\zeta - z} \left(\int_{\partial S} \frac{f(\zeta, \eta)}{\eta - \zeta} \, d\eta \right) d\zeta = -\pi^2 f(z, z) + \int_{\partial S} \left(\int_{\partial S} \frac{f(\zeta, \eta)}{(\zeta - z)(\eta - \zeta)} \, d\zeta \right) d\eta. \tag{A66}$$

Using (A66) and the fact that, by (A64) with $\varphi = 1$,

$$\int_{\partial S} \frac{d\zeta}{\zeta - z} = \pi i, \quad z \in \partial S, \tag{A67}$$

in the sense of principal value, we find that for any $\varphi \in C^{0,\alpha}(\partial S)$ and $z \in \partial S$

$$((K^s)^2 \varphi)(z) = \int_{\partial S} \frac{1}{\zeta - z} \left(\int_{\partial S} \frac{\varphi(\eta)}{\eta - \zeta} \, d\eta \right) d\zeta = -\pi^2 \varphi(z) + \int_{\partial S} \left(\int_{\partial S} \frac{\varphi(\eta)}{(\zeta - z)(\eta - \zeta)} \, d\zeta \right) d\eta$$

$$= -\pi^2 \varphi(z) + \int_{\partial S} \left[\frac{1}{\eta - z} \left(\int_{\partial S} \frac{d\zeta}{\zeta - z} - \int_{\partial S} \frac{d\zeta}{\zeta - \eta} \right) \varphi(\eta) \right] d\eta = -\pi^2 \varphi(z),$$

as required.

A65. Theorem. *Let $f(z, \zeta)$ be a function defined on $\partial S \times \partial S$ which belongs to $C^{0,\alpha}(\partial S)$, $\alpha \in (0,1]$, with respect to each of its variables, uniformly relative to the other one, and satisfies the inequality*

$$|f(z, \zeta) - f(z', \zeta)| < c|z - z'| \, |z - \zeta|^{\alpha - 1}, \quad c = \text{const} > 0,$$

for all $z, z', \zeta \in \partial S$ such that $0 < |z - z'| < \frac{1}{2}|z - \zeta|$. Then the function

$$\Lambda(z) = \int_{\partial S} \frac{f(z,\zeta)}{\zeta - z}\, d\zeta, \quad z \in \partial S,$$

where the integral is understood as principal value, belongs to $C^{0,\beta}(\partial S)$, with $\beta = \alpha$ for $\alpha \in (0,1)$ and any $\beta \in (0,1)$ for $\alpha = 1$.

Proof. Let $z = x_1 + ix_2$. Writing

$$\int_{\partial S \setminus \Sigma_{x,r}} \frac{f(z,\zeta)}{\zeta - z}\, d\zeta = \int_{\partial S \setminus \Sigma_{x,r}} \frac{f(z,\zeta) - f(z,z)}{\zeta - z}\, d\zeta + f(z,z) \int_{\partial S \setminus \Sigma_{x,r}} \frac{d\zeta}{\zeta - z},$$

from Theorem A61 and the fact that the integrand of the first term on the right-hand side is $O(|z - \zeta|^{\alpha - 1})$, we conclude that $\Lambda(z)$ exists in the sense of principal value for all $z \in \partial S$.

To establish the Hölder continuity of Λ, for $z, z', \zeta \in \partial S$ we use the decomposition

$$2[\Lambda(z) - \Lambda(z')] = \int_{\partial S} \left(\frac{f(z,\zeta) - f(z,z)}{\zeta - z} - \frac{f(z,\zeta) - f(z,z')}{\zeta - z'} \right) d\zeta$$

$$+ \int_{\partial S} \left(\frac{f(z',\zeta) - f(z',z)}{\zeta - z} - \frac{f(z',\zeta) - f(z',z')}{\zeta - z'} \right) d\zeta$$

$$+ \int_{\partial S} \frac{f(z,\zeta) - f(z',\zeta)}{\zeta - z}\, d\zeta + \int_{\partial S} \frac{f(z,\zeta) - f(z',\zeta)}{\zeta - z'}\, d\zeta$$

$$+ f(z,z) \int_{\partial S} \frac{d\zeta}{\zeta - z} - f(z,z') \int_{\partial S} \frac{d\zeta}{\zeta - z'}$$

$$+ f(z',z) \int_{\partial S} \frac{d\zeta}{\zeta - z} - f(z',z') \int_{\partial S} \frac{d\zeta}{\zeta - z'}$$

$$= I_1 + I_2 + I_3 + I_4 + I_5 + I_6.$$

Let $z' = x_1' + ix_2'$ and $\zeta = y_1 + iy_2$, and let $\Sigma_{x,r}$, Σ_1 and Σ_2 be the sets defined by (A6), (A9) and (A10), with x and x' satisfying (A12). By Lemmas A8–A11, (A67) and Remark A36,

$$|I_{11}| = \left| \int_{\Sigma_1} \left(\frac{f(z,\zeta) - f(z,z)}{\zeta - z} - \frac{f(z,\zeta) - f(z,z')}{\zeta - z'} \right) d\zeta \right|$$

$$\leq c_1 \int_{\Sigma_1} (|x - y|^{\alpha - 1} + |x' - y|^{\alpha - 1})\, ds(y) \leq c_2 |z - z'|^\alpha,$$

$$|I_{12}| = \left| \int_{\Sigma_2} \left(\frac{1}{\zeta - z} - \frac{1}{\zeta - z'} \right) [f(z, \zeta) - f(z, z')] \, d\zeta \right|$$

$$\leq c_3 |z - z'| \int_{\Sigma_2} |x - y|^{\alpha - 2} ds(y) \leq c_4 |z - z'|^{\alpha} \quad \text{if } \alpha \in (0, 1),$$

$$|I_{12}| \leq c_5 |z - z'|^{\alpha} |\ln |z - z'|| \quad \text{if } \alpha = 1,$$

$$|I_{13}| = \left| \int_{\partial S \setminus \Sigma_{x,r}} \left(\frac{1}{\zeta - z} - \frac{1}{\zeta - z'} \right) [f(z, \zeta) - f(z, z')] \, d\zeta \right|$$

$$\leq c_6 |z - z'| \int_{\partial S \setminus \Sigma_{x,r}} |x - y|^{\alpha - 2} ds(y) \leq c_7 |z - z'|,$$

$$|I_{14}| = \left| [f(z, z') - f(z, z)] \int_{\partial S \setminus \Sigma_1} \frac{d\zeta}{\zeta - z} \right| \leq c_8 |z - z'|^{\alpha};$$

consequently,

$$|I_1| = |I_{11} + I_{12} + I_{13} + I_{14}| \leq c_9 |z - z'|^{\beta},$$

where the constants $c_1, \ldots, c_9 > 0$ may depend on α.

Similarly,

$$|I_2| \leq c_{10} |z - z'|^{\beta}, \quad c_{10} = \text{const} > 0.$$

Next, we find that

$$|I_{31}| = \left| \int_{\Sigma_1} \{ [f(z, \zeta) - f(z, z)] - [f(z', \zeta) - f(z', z)] \} \frac{d\zeta}{\zeta - z} \right|$$

$$\leq c_{11} \int_{\Sigma_1} |x - y|^{\alpha - 1} ds(y) \leq c_{12} |z - z'|^{\alpha},$$

$$|I_{32}| = \left| \int_{\Sigma_2} \frac{f(z, \zeta) - f(z', \zeta)}{\zeta - z} \, d\zeta \right|$$

$$\leq c_{13} |z - z'| \int_{\Sigma_2} |x - y|^{\alpha - 2} ds(y) \leq c_{14} |z - z'|^{\alpha} \quad \text{if } \alpha \in (0, 1),$$

$$|I_{32}| \leq c_{15} |z - z'| |\ln |z - z'|| \quad \text{if } \alpha = 1,$$

$$|I_{33}| = \left| \int_{\partial S \setminus \Sigma_{x,r}} \frac{f(z, \zeta) - f(z', \zeta)}{\zeta - z} \, d\zeta \right|$$

$$\leq c_{16} |z - z'| \int_{\partial S \setminus \Sigma_{x,r}} |x - y|^{\alpha - 2} ds(y) \leq c_{17} |z - z'|,$$

$$|I_{34}| = \left|[f(z, z) - f(z', z)]\left(\int_{\partial S} \frac{d\zeta}{\zeta - z} - \int_{\partial S \setminus \Sigma_1} \frac{d\zeta}{\zeta - z}\right)\right| \leq c_{18}|z - z'|^{\alpha};$$

therefore,
$$|I_3| = |I_{31} + I_{32} + I_{33} + I_{34}| \leq c_{19}|z - z'|^{\beta},$$

where the constants $c_{11}, \ldots, c_{19} > 0$ may depend on α. In exactly the same way, but using $\Sigma_{x',r}$ instead of $\Sigma_{x,r}$, we find that

$$|I_4| \leq c_{20}|z - z'|^{\beta}, \quad c_{20} = \text{const} > 0.$$

Finally,
$$|I_5| = \left|\pi i[f(z, z) - f(z, z')]\right| \leq c_{21}|z - z'|^{\alpha},$$
$$|I_6| = \left|\pi i[f(z', z) - f(z', z')]\right| \leq c_{22}|z - z'|^{\alpha},$$

where c_{21} and c_{22} are positive constants.

Combining the above inequalities, we now obtain
$$|\Lambda(z) - \Lambda(z')| \leq c_{23}|z - z'|^{\beta}, \quad c_{23} = \text{const} > 0,$$

as required.

A7. Singular integral equations

We discuss briefly the elements of functional analysis mentioned in §1.1.

A66. Theorem. $C^{0,\alpha}(\partial S)$ is a Banach space with respect to the metric defined in Theorem 1.5.

Proof. It is easy to verify that the mapping

$$\|\varphi\|_\alpha = \|\varphi\|_\infty + |\varphi|_\alpha \tag{A68}$$

where
$$\|\varphi\|_\infty = \sup_{z \in \partial S} |\varphi(z)|, \quad |\varphi|_\alpha = \sup_{\substack{z,\zeta \in \partial S \\ z \neq \zeta}} \frac{|\varphi(z) - \varphi(\zeta)|}{|z - \zeta|^\alpha},$$

satisfies the norm axioms.

Let $\{\varphi_n\}_{n=1}^\infty$ be a Cauchy sequence in $C^{0,\alpha}(\partial S)$; that is, for any $\varepsilon > 0$ arbitrarily small there is a positive integer $n_0(\varepsilon)$ such that

$$\|\varphi_n - \varphi_m\|_\alpha < \varepsilon \quad \text{for all } n, m > n_0(\varepsilon).$$

By (A68),
$$\|\varphi_n - \varphi_m\|_\infty < \varepsilon \quad \text{for all } n, m > n_0(\varepsilon),$$
which means that $\{\varphi_n\}_{n=1}^\infty$ is also a Cauchy sequence in $C(\partial S)$. Since $C(\partial S)$ is a complete space, there is $\varphi \in C(\partial S)$ such that
$$\|\varphi_n - \varphi\|_\infty \to 0 \quad \text{as } n \to \infty. \tag{A69}$$

From (A68) we also deduce that
$$|\varphi_n - \varphi_m|_\alpha < \varepsilon \quad \text{for all } n, m > n_0(\varepsilon).$$

Passing to the limit as $m \to \infty$ and using the uniform convergence of $\{\varphi_n\}_{n=1}^\infty$ on ∂S, we now obtain
$$|\varphi_n - \varphi|_\alpha < \varepsilon \quad \text{for all } n > n_0(\varepsilon). \tag{A70}$$

Hence, there is $c = \text{const} > 0$ such that for all $z, \zeta \in \partial S$, $z \neq \zeta$,
$$\frac{|\varphi(z) - \varphi(\zeta)|}{|z - \zeta|^\alpha} \leq |\varphi|_\alpha \leq c;$$

in other words, $\varphi \in C^{0,\alpha}(\partial S)$. Also, from (A69) and (A70) it follows that
$$\|\varphi_n - \varphi\|_\alpha \to 0 \quad \text{as } n \to \infty;$$

that is, $\{\varphi_n\}_{n=1}^\infty$ converges in the norm (A68), which means that $C^{0,\alpha}(\partial S)$ is complete.

A67. Theorem. *If $k(z, \zeta)$ is a proper γ-singular kernel on ∂S, $\gamma \in [0, 1)$, then the operator K defined by*
$$(K\varphi)(z) = \int_{\partial S} k(z, \zeta)\varphi(\zeta)\, d\zeta, \quad z \in \partial S, \tag{A71}$$

is a compact operator from $C^{0,\alpha}(\partial S)$ into $C^{0,\alpha}(\partial S)$, with $\alpha = 1 - \gamma$ for $\gamma \in (0, 1)$ and any $\alpha \in (0, 1)$ for $\gamma = 0$.

Proof. According to Theorem A28 and the fact that $C^{0,\alpha}(\partial S) \subset C(\partial S)$, the operator $K : C^{0,\alpha}(\partial S) \to C^{0,\alpha}(\partial S)$ is well defined.

Let $M_1 \subset C^{0,\alpha}(\partial S)$ be a bounded set, that is,
$$\|\varphi\|_\alpha \leq c = \text{const} > 0 \quad \text{for all } \varphi \in M_1. \tag{A72}$$

Also, let $\{\theta_n\}_{n=1}^{\infty} \subset M_2 = K(M_1)$. We denote by $\{\varphi_n\}_{n=1}^{\infty}$ a sequence in M_1 such that $\theta_n = K\varphi_n$, $n = 1, 2, \ldots$

In view of (A68), inequality (A72) implies that

$$\sup_{z \in \partial S} |\varphi_n(z)| \leq c,$$

$$|\varphi_n(z) - \varphi_n(z')| \leq c|z - z'|^\alpha$$

for all $n = 1, 2, \ldots$ and all x, $x' \in \partial S$; in other words, $\{\varphi_n\}_{n=1}^{\infty}$ is uniformly bounded and equicontinuous in $C(\partial S)$. By the Arzelà-Ascoli Theorem, it contains a uniformly convergent subsequence. For simplicity, we denote this subsequence again by $\{\varphi_n\}_{n=1}^{\infty}$. Hence, there is a $\varphi \in C(\partial S)$ such that

$$\|\varphi_n - \varphi\|_\infty \to 0 \quad \text{as } n \to \infty. \tag{A73}$$

Let $\theta = K\varphi$. By Theorem A28, $\theta \in C^{0,\alpha}(\partial S)$. For $z \in \partial S$ we have

$$|\theta_n(z) - \theta(z)| \leq \int_{\partial S} |k(z, \zeta)| |\varphi_n(\zeta) - \varphi(\zeta)| d\zeta$$

$$\leq c_1 \sup_{x \in \partial S} |\varphi_n(z) - \varphi(z)| \int_{\partial S} |z - \zeta|^{-\gamma} d\zeta;$$

consequently, by Theorem A27,

$$\|\theta_n - \theta\|_\infty \leq c_2 \|\varphi_n - \varphi\|_\infty, \quad n = 1, 2, \ldots,$$

where c_1 and c_2 are positive constants. On the other hand, by Theorem A28,

$$|\theta_n - \theta|_\alpha \leq c_3 \|\varphi_n - \varphi\|_\alpha, \quad n = 1, 2, \ldots$$

The last two inequalities together with (A68) and (A73) yield

$$\|\theta_n - \theta\|_\alpha \to 0 \quad \text{as } n \to \infty,$$

which proves that K is compact on $C^{0,\alpha}(\partial S)$.

A68. Remark. Let K be the operator defined by (A71), and consider the dual system $\bigl(C^{0,\alpha}(\partial S), C^{0,\alpha}(\partial S)\bigr)$, $\alpha \in (0, 1)$, with the bilinear form

$$(\varphi, \psi) = \int_{\partial S} \varphi(\zeta)\psi(\zeta) d\zeta, \quad \varphi, \psi \in C^{0,\alpha}(\partial S), \tag{A74}$$

which is easily seen to be non-degenerate (according to Definition 1.9). Using (A74), we find that the operator defined on $C^{0,\alpha}(\partial S)$ by

$$(K\varphi, \psi) = \int_{\partial S} \left(\int_{\partial S} k(z,\zeta)\varphi(\zeta)\,d\zeta \right) \psi(z)\,dz$$
$$= \int_{\partial S} \varphi(\zeta) \left(\int_{\partial S} k(z,\zeta)\psi(z)\,dz \right) d\zeta = (\varphi, K^*\psi),$$

where
$$(K^*\varphi)(z) = \int_{\partial S} k^*(z,\zeta)\varphi(\zeta)\,d\zeta$$

with $k^*(z,\zeta) = k(\zeta, z)$, is the (unique) adjoint of K, as mentioned in Definition 1.11. This means that if $k(z,\zeta)$ is a proper $(1-\alpha)$-singular kernel on ∂S with respect to both z and ζ, then, by Theorems A67 and 1.14, the Fredholm Alternative holds for K and K^*.

Theorem A65 enables us to introduce the concept of α-*regular singular operator* (see Definition 1.16).

A69. Theorem. *If $k(x, y)$ is a proper γ-singular kernel on ∂S, $\gamma \in [0, 1)$, with respect to both x and y, then the operator K defined on $C^{0,1-\gamma}(\partial S)$ by*

$$(K\varphi)(x) = \int_{\partial S} k(x, y)\varphi(y)\,ds(y), \quad x \in \partial S,$$

is $(1-\gamma)$-regular singular, and the function $\hat{k}(z,\zeta)$ in the expression of K in Definition 1.16 *satisfies*
$$\hat{k}(z, z) = 0, \quad z \in \partial S.$$

Proof. Clearly, $(K\varphi)(x)$ is an improper integral for all $x \in \partial S$.

In accordance with our notational convention, we write

$$\int_{\partial S} k(x,y)\varphi(y)\,ds(y) = \int_{\partial S} k(z,\zeta)\varphi(\zeta)\bar{\vartheta}(\zeta)\,d\zeta = \int_{\partial S} \frac{\hat{k}(z,\zeta)}{\zeta - z}\,d\zeta,$$

where $\vartheta(z)$ is defined by (A61) and
$$\hat{k}(z,\zeta) = (\zeta - z)k(z,\zeta)\bar{\vartheta}(\zeta).$$

By Definition 1.2, $\hat{k}(z,z) = 0$ in the sense of continuous extension.

Let $z, z', \zeta \in \partial S$ be such that $0 < |z - z'| < \frac{1}{2}|z - \zeta|$. In this case,

$$|z' - \zeta| \geq |z - \zeta| - |z - z'| > |z - \zeta| - \tfrac{1}{2}|z - \zeta| = \tfrac{1}{2}|z - \zeta|. \tag{A75}$$

Since $|\vartheta(\zeta)| = 1$, $\zeta \in \partial S$, we have

$$\begin{aligned}|\hat{k}(z,\zeta) - \hat{k}(z',\zeta)| &= |(\zeta - z)k(z,\zeta) - (\zeta - z')k(z',\zeta)| \\ &\leq |z - \zeta|\,|k(z,\zeta) - k(z',\zeta)| + |z - z'|\,|k(z',\zeta)| \\ &\leq c_1|z - z'|(|z - \zeta|^{-\gamma} + |z' - \zeta|^{-\gamma}),\end{aligned}$$

which, on the basis of (A75), shows that

$$|\hat{k}(z,\zeta) - \hat{k}(z',\zeta)| \leq c_2|z - z'|\,|z - \zeta|^{-\gamma}$$

and that

$$|\hat{k}(z,\zeta) - \hat{k}(z',\zeta)| \leq c_3|z - z'|^{1-\gamma},$$

where the constants $c_1, c_2, c_3 > 0$ are independent of z, z' and ζ.
If $|z - z'| \geq \frac{1}{2}|z - \zeta|$, then

$$|z' - \zeta| \leq |z - z'| + |z - \zeta| \leq 3|z - z'|;$$

consequently,

$$\begin{aligned}|\hat{k}(z,\zeta) - \hat{k}(z',\zeta)| &\leq |z - \zeta|\,|k(z,\zeta)| + |z' - \zeta|\,|k(z',\zeta)| \\ &\leq c_1(|z - \zeta|^{1-\gamma} + |z' - \zeta|^{1-\gamma}) \leq c_4|z - z'|^{1-\gamma},\end{aligned}$$

where c_4 is independent of z, z' and ζ.

The Hölder continuity of $\hat{k}(z,\zeta)$ with respect to ζ is proved similarly: we write

$$\begin{aligned}|\hat{k}(z,\zeta) - \hat{k}(z,\zeta')| &\leq |\zeta - z|\,|k(z,\zeta)|\,|\vartheta(\zeta) - \vartheta(\zeta')| \\ &\quad + |\zeta - z|\,|k(z,\zeta) - k(z,\zeta')| + |\zeta - \zeta'|\,|k(z,\zeta')|\end{aligned}$$

and use the fact that $\vartheta \in C^{0,\alpha}(\partial S)$.

The importance of α-regular singular operators is illustrated by Theorem 1.19. If the index (see Definition 1.17) of such an operator is zero, then the Fredholm Alternative holds for it in the dual system $\bigl(C^{0,\alpha}(\partial S), C^{0,\alpha}(\partial S)\bigr)$ with the non-degenerate bilinear form (A74). A comprehensive discussion of this assertion can be found in [19] and [20].

References

1. M. Abramowitz and I. Stegun, *Handbook of mathematical functions*, Dover, New York, 1964.
2. D. Colton and R. Kress, *Integral equation methods in scattering theory*, Wiley, New York, 1983.
3. C. Constanda, *A mathematical analysis of bending of plates with transverse shear deformation*, Pitman Res. Notes Math. Ser. **215**, Longman, Harlow-New York, 1990.
4. C. Constanda, On the solution of the Dirichlet problem for the two-dimensional Laplace equation, *Proc. Amer. Math. Soc.* **119** (1993), 877–884.
5. C. Constanda, Sur le problème de Dirichlet dans la déformation plane, *C.R. Acad. Sci. Paris Sér. I* **316** (1993), 1107–1109.
6. C. Constanda, On non-unique solutions of weakly singular integral equations in plane elasticity, *Quart. J. Mech. Appl. Math.* **47** (1994), 261–268.
7. C. Constanda, On integral solutions of the equations of thin plates, *Proc. Roy. Soc. London Ser. A* **444** (1994), 317–323.
8. C. Constanda, The boundary integral equation method in plane elasticity, *Proc. Amer. Math. Soc.* **123** (1995), 3385–3396.
9. C. Constanda, Integral equations of the first kind in plane elasticity, *Quart. Appl. Math.* **53** (1995), 783–793.
10. C. Constanda, On the direct and indirect methods in the theory of elastic plates, *Math. Mech. Solids* **1** (1996), 251–260.
11. C. Constanda, Unique solution in the theory of elastic plates, *C.R. Acad. Sci. Paris Sér. I* **323** (1996), 95–99.
12. C. Constanda, Robin-type conditions in plane strain, in *Integral methods in science and engineering, vol. 1: analytic techniques*, Pitman Res. Notes Math. Ser. **374**, Addison Wesley Longman, Harlow-New York, 1997, 55–59.
13. C. Constanda, Fredholm equations of the first kind in the theory of bending of elastic plates, *Quart. J. Mech. Appl. Math.* **50** (1997), 85–96.
14. C. Constanda, On boundary value problems associated with Newton's law of cooling, *Appl. Math. Lett.* **10** (1997), no. 5, 55–59.
15. C. Constanda, Elastic boundary conditions in the theory of plates, *Math. Mech. Solids* **2** (1997), 189–197.

16. N.M. Günter, *Potential theory and its applications to basic problems of mathematical physics*, F. Ungar, New York, 1967.
17. E. Hille, *Analytic function theory*, Ginn, Boston-London, 1959.
18. M.A. Jaswon and G.T. Symm, *Integral equation methods in potential theory and elastostatics*, Academic Press, London, 1977.
19. V.D. Kupradze et al., *Three-dimensional problems of the mathematical theory of elasticity and thermoelasticity*, North-Holland, Amsterdam, 1979.
20. N.I. Muskhelishvili, *Singular integral equations*, P. Noordhoff, Groningen, 1946.
21. P. Schiavone, On the Robin problem for the equations of thin plates, *J. Integral Equations Appl.* **8** (1996), 231–238.
22. V.I. Smirnov, *A course of higher mathematics*, vol. 2, Pergamon Press, Oxford, 1964.
23. G.R. Thomson and C. Constanda, Area potentials for thin plates, *An. Stiint. Univ. Al. I. Cuza Iasi Sect. Ia Mat.* **44** (1998), no. 2.
24. G.R. Thomson and C. Constanda, Smoothness properties of a domain potential arising in the study of elastic plates, *Strathclyde Math. Res. Report* **11** (1998).
25. V.S. Vladimirov, *Equations of mathematical physics*, M. Dekker, New York, 1971.